The Essential Human Anatomy Compendium (Second Edition)

A Comprehensive and Concise Study Guide for Success in Introductory Anatomy Courses

Professor H.P. Doyle

authorHOUSE®

AuthorHouse™
1663 Liberty Drive
Bloomington, IN 47403
www.authorhouse.com
Phone: 1-800-839-8640

First published by AuthorHouse 7/7/2009

Printed in the United States of America
Bloomington, Indiana

This book is printed on acid-free paper.

ISBN: 978-1-4389-8648-7 (sc)

PREFACE

Welcome to the **Second Edition** of The Essential Human Anatomy Compendium, which is a **study guide in the format of LECTURE OUTLINE NOTES compiled from various university instructors nationwide**. Students have expressed that using this essential study guide is a major positive step toward excelling in their college-level (or advanced high school level) Human Anatomy course.

How is our study guide different from others already in publication?

The format of this book is the outline form, which **lends itself to easy perusing**. KEY WORDS or PHRASES are EMPHASIZED VISUALLY and as CONCISELY as possible, in order to **break up the monotony**, which is often seen in long-winded textbooks. Though the goal is **brevity**, these outline notes still provide COPIOUS INFORMATION, which is *not* represented in other study guides in existence. The approach of this study guide is to allow the student to **comprehend the gist of basic anatomical concepts**.

This study guide is organized into five key sections:
(1) **Introductory and Microscopic Anatomy**, including cytology (cell study) and histology (tissue study); (2) **Skeletal Anatomy**, including axial and appendicular skeletal anatomy and accessory structures; (3) **Muscular Anatomy**, focusing on the origin, insertion, and action of key muscles required for most students to learn; (4) **Neuroanatomy**, including the Central Nervous System (brain and spinal cord), Peripheral Nervous System (including critical Autonomic Nervous System features), and general and special Sensory Anatomy; and (5) **Systemic Anatomy**, targeting the eleven human body organ systems and their components.

This latest edition includes 50% MORE multiple-choice questions for practice reviews, which will prepare you for the key levels of anatomy exam questions. These questions have been developed by various instructors from several disciplines.

For Instructors: Answer sheets to the questions are also provided after each set of questions so that students may complete them and submit them for instructor review (and perhaps for credit).

How to use *The Essential Human Anatomy Compendium*:

Due to the nature of this book, it should be utilized as a **key study tool** prior to course exams, prior to, after and/or during class lectures, or it may be used as a **remedial preparation tool** for Board exams in various disciplines. **We have intentionally excluded figures from this study guide to compel students to use it as a primary reference that is best used in conjunction with the assigned textbook.**

Whether your academic training specialty is in Nursing, Dentistry, Dental Hygiene, Occupational or Physical Therapy, Athletic Training, Exercise Science, Pharmacy, or other Allied Health disciplines, you will undoubtedly find *The Essential Human Anatomy Compendium* a useful tool, which will help you **excel in the subject of anatomy**.

Good luck on your journey of discovery!

H.P. Doyle

ABBREVIATIONS KEY and STUDY TOOLS

Throughout this study guide, there are *Study Tips*, *Mnemonic Devices*, and *Memory Aid* **suggestion boxes**, which are mainly memory tricks/advice to facilitate easy learning and/or to reinforce and clarify new terminologies or concepts.

Study Tips: These are hints or suggestions to help the student learn and retain a concept or terminology.

Mnemonic device: a memory aid; often verbal, such as a very short poem / phrase or a special word used to help a person remember something, <u>particularly lists</u>, but may be visual, kinesthetic or auditory

Acronym: abbreviation that is formed using the initial components in a phrase or name; often used in scientific vernacular: e.g. CNS = Central Nervous System

e.g. = "*exempli gratia*" – 'for the sake of example' or '<u>for example</u>'

i.e. = "*id est*" - "That is (to say)", "<u>in other words</u>", or sometimes "<u>in this case</u>", depending on the context. Never equivalent to exempli gratia (e.g.), although frequently so used.

n.b. = "*nota bene*" - That is, "please note" or "note it well".

pl. = plural form of a term

sing. = singular form of a term

BRIEF TABLE OF CONTENTS

Chapter 1 – INTRODUCTION to ANATOMY of the HUMAN BODY

ANATOMY (also called *morphology* – the science of form) is the study of internal and external structures of the human body. Given that specific structures perform specific functions, therefore, **structure determines function**.

The BRANCHES of ANATOMY include:

- **MICROSCOPIC ANATOMY** (fine anatomy) – the study of body structures that cannot be viewed without magnification; includes <u>cytology</u> (the study of cells) and <u>histology</u> (the study of tissues)

- **GROSS ANATOMY** (macroscopic anatomy) – the study of body structures which are visible without the aid of magnification; subclasses include <u>surface anatomy</u> (the study of shapes and markings on the body surface), <u>regional anatomy</u> (the study of all structures in a single body region, superficial or deep) and <u>systemic anatomy</u> (the study of all organs with related functions, i.e. study one organ system at a time)

- **DEVELOPMENTAL ANATOMY** – the study of structural changes that occur from conception to physical maturity; its subclass is <u>embryology</u> (the study of structural formation and development before birth)

- **COMPARATIVE ANATOMY** – the study of the anatomy of different types of animals

- **PATHOLOGICAL ANATOMY** – the study of structural changes in cells tissues, and organs caused by disease

- **RADIOGRAPHIC ANATOMY** – the study of internal body structures by using noninvasive imaging techniques, such as X-ray imaging and ultrasound

- **SURGICAL ANATOMY** - the study of anatomical landmarks, which are important to surgical procedures

A. **LEVELS of STRUCTURAL ORGANIZATION of the human body** (*microscopic to macroscopic*):

1. **CHEMICAL LEVEL**: *atoms* (building blocks of matter) combine to form small *molecules* (such as water and carbon dioxide) and larger *macromolecules* (carbohydrates, lipids, proteins and nucleic acids); i.e., Chemicals (molecules) comprise the entire body.

2. **CELLULAR LEVEL**: cells are comprised of molecules; they are the smallest living units in the body; cellular <u>organelles</u> are their functional subunits.

3. **TISSUE LEVEL**: similar types of cells, with a *common function*, combine to form tissues; four primary tissue types comprise all organs of the human body.

4. **ORGAN LEVEL**: more than one tissue type (often all four tissues) combine to form organs; extremely complex physiological processes occur at this level.

5. **ORGAN SYSTEM LEVEL**: organs that work closely together combine to form an organ system, to accomplish a common purpose; there are 11 organ systems of the human body.

6. **HUMAN ORGANISM** (or organismal level): the highest level of structural organization; this is the combination of all the organ systems functioning together to sustain the life of the organism.

B. **The HUMAN BODY'S 11 ORGAN SYSTEMS**:

1. **Integumentary System** - forms the cutaneous membrane (epidermis and dermis), the external body covering; <u>provides protection and thermoregulation</u>; synthesizes vitamin D; provides cutaneous reception through sensory receptors; additional accessory structures are hair follicles, nails, sweat (sudoriferous) glands and oil (sebaceous) glands

2. **Skeletal System** – provides protection and support to the body organs; provides skeletal framework for the muscles to attach, hence, causing movement; stores minerals; blood cell formation occurs within bones

3. **Muscular System** – produces motion; maintains posture by providing support; produces heat

4. **Nervous System** – control center of the body, which directs immediate responses to stimuli and coordinates the other organ systems; i.e. responds to internal and external stimuli by activating appropriate muscles and glands

5. **Endocrine System** – comprised of glands, which secrete hormones that regulate processes (such as metabolism, growth, and reproduction) of the other organ systems

6. **Cardiovascular System** – comprised of the heart, blood vessels, and blood to transport materials (such as respiratory gases, nutrients and wastes) within the body

7. **Lymphoid (Lymphatic and Immune) System** – comprised of the lymphatic vessels, lymphoid organs (lymph nodes, the thymus, and the spleen), lymphocytes, and lymphoid tissue; returns leaked fluid to blood; provides defense against pathogens and disease by housing white bloods cells (lymphocytes) that function in immunity

8. **Respiratory System** – comprised of the nasal cavities, pharynx, larynx, trachea, bronchi, and the lungs; maintains the blood's constant supply of oxygen and removal of carbon dioxide by delivering air to the lungs where gas exchange occurs at the alveoli (air sacs of the lungs)

9. **Digestive System** – comprised of the gastrointestinal tract (or alimentary canal) and accessory structures, which together function to process food and absorb nutrients

10. **Urinary System** – comprised of the kidneys, ureters, the urinary bladder, and the urethra; functions to eliminate excess water, salts, and nitrogenous wastes from the body; regulates water, electrolyte, and acid-base balance of the blood

11. **Reproductive System** – comprised of gonads (testes in the male; ovaries in the female), accessory organs and external genitalia; overall function is to produce sex cells and hormones for the purpose of producing offspring; the female reproductive system supports embryonic development

C. ANATOMICAL POSITON: AXIAL vs. APPENDICULAR REGIONS

- In **anatomical position**, the person is standing upright, arms at sides, palms facing forward (little fingers are medial, touching the thighs), feet flat on the floor, face straight ahead.

    - The **AXIAL REGION** consists of the head, neck and torso.

    - The **APPENDICULAR REGION** consists of the upper and lower limbs (or extremities).

D. ORIENTATION and DIRECTIONAL TERMS:

- **SUPERIOR (cranial or cephalic)** – above the point of reference; toward the head end or upper part of a structure or the body

- **INFERIOR (caudal)** – below the point of reference; toward the tail end or toward the lower part of a structure or the body

- **ANTERIOR (ventral)** – toward or at the front of the body; in front of; (the front or belly side)

- **POSTERIOR (dorsal)** – toward or at the back of the body; behind; (the back side) (**n.b.** anterior/ dorsal and posterior/ventral are interchangeable in humans only, not so in four-footed animals in which dorsal is superior and ventral is inferior)

- **MEDIAL** – toward or at the midline of the body; on the inner side of (e.g. The trachea is medial to the arm.)

- **LATERAL** – away from the midline of the body; toward the sides or on the outer sides of (e.g. The ears are lateral to the nose.)

- **PROXIMAL** – closer to the origin of the body part or the point of attachment of a limb to the body trunk (e.g. The shoulders are proximal to the elbows.)

- **DISTAL** – away from the origin of a body part or the point of attachment of a limb to the body trunk (e.g. The fingers are distal to the wrist.)

- **SUPERFICIAL (external)** – closer to or at the body surface (e.g. The skin is superficial to the heart.)

- **DEEP (internal)** – farther from or away from the body surface; more internal (e.g. The lungs are deep to the skin.)

- **IPSILATERAL** – on the same side (e.g. The right arm and right leg are ipsilateral.)

- **CONTRALATERAL** – on <u>opposite sides</u> (e.g. The right arm and left leg are contralateral.)

E. REGIONAL TERMS (names of specific body parts)
The following is a list of the most commonly used anatomical terms.

<u>Anatomical NAME (*Anatomical REGION*) – Common Term:</u>
- CEPHALON (*cephalic*) – area of the head
- CERVICIS (*cervical*) – neck region
- THORACIS (*thoracic*) – chest region
- BRACHIUM (*brachial*) – upper arm
- ANTEBRACHIUM (*antebrachial*) - forearm
- CARPUS (*carpal*) – wrist
- MANUS (*manual*) – hand
- POLLICIS (*pollex*) - thumb
- ABDOMEN (*abdominal*) – abdominal region
- UMBILICUS (*umbilical*) – navel or bellybutton
- PELVIS (*pelvic*) – pelvic region
- PUBIS (*pubic*) – anterior pelvis or genital region
- INGUEN (*inguinal*) – groin
- LUMBUS (*lumbar*) – lower back
- GLUTEUS (*gluteal*) – buttock region
- FEMUR (*femoral*) – thigh
- PATELLA (*patellar*) – kneecap
- CRUS (*crural*) – anterior leg, from knee to ankle
- SURA (*sural*) – posterior, calf of leg
- TARSUS (*tarsal*) – ankle
- PES (*pedal*) – foot (pedals of a bike)
- PLANTA (*plantar*) – the bottom of the foot, sole
- HALLUCIS (*hallux*) – great toe or big toe

F. BODY PLANES

- **FRONTAL (CORONAL) plane** – lies vertically and divides the body into an anterior (front) portion and a posterior (back) portion

- **TRANSVERSE (HORIZONTAL) plane** – lies horizontally and divides the body into a superior (top) portion and an inferior (lower) portion. These sections are also called **cross sections**.

- **SAGITTAL plane** – lies vertically and divides the body into a right portion and a left portion. If the sagittal plane lies exactly in the midline and the <u>portions are equivalent</u>, it is called a **MIDSAGITTAL** PLANE**, or **MEDIAN** PLANE; all other sagittal planes (that are offset from the midline and result in <u>unequal portions</u>) are called **PARASAGITTAL** PLANES.

G. BODY CAVITIES and MEMBRANES

1. **DORSAL body cavity**
   a. <u>Cranial cavity</u> – lies within skull (cranium), encasing the brain
   b. <u>Spinal cavity</u> – lies within the vertebral column, enclosing the spinal cord

2. **VENTRAL body cavity (COELOM)** – provides protection, allows organ movement, lining prevents friction

   a. **The THORACIC cavity** – superior to diaphragm, contains heart, lungs, blood vessels; surrounded by the ribs and the muscles of the chest wall

      o **PLEURAL CAVITIES** – right and left cavities, which enclose the right and left lungs
      - <u>parietal pleura</u> (the thin membrane that lines the chest walls) of the serous membrane

      - <u>visceral pleura</u> (the thin membrane that adheres to the lungs) of the serous membrane

      - <u>serous fluid</u> fills the pleural cavity between the layers of the serous membrane

      o **MEDIASTINAL CAVITY or MEDIASTINUM** – a central cavity containing a band of organs, which lies between the pleural cavities; contains the heart (enclosed by the pericardial cavity), esophagus, trachea, and major blood vessels
      - **PERICARDIAL CAVITY** – contains the heart

         - <u>parietal pericardium</u> (the thin membrane that lines the pericardial walls) of the serous membrane

         - <u>visceral pericardium</u> (the thin membrane that adheres to the heart surface) of the serous membrane

         - <u>serous fluid</u> fills the pericardial cavity between the layers of the serous membrane

   b. **The ABDOMINOPELVIC CAVITY** – lies inferior to the diaphragm and is divided into a superior part and an inferior part

      o **ABDOMINAL cavity** – the superior part, which contains the liver, stomach, small intestine, spleen, kidneys, and other organs; extends from diaphragm superiorly to superior border of sacrum

      - Many organs in the abdominopelvic cavity are surrounded by a peritoneal cavity.

      - **peritoneum** – serous membrane

         - <u>parietal peritoneum</u> (the thin membrane that lines the wall) of the serous membrane

- visceral peritoneum (the thin membrane that adheres to the abdominopelvic organs) of the serous membrane

  - Note that the kidneys, adrenal glands, pancreas, and ureters are **retroperitoneal** because they are located behind the abdominopelvic cavity.

  o **PELVIC cavity** – the inferior part, which is enclosed by the bony pelvis; contains the urinary bladder, some reproductive organs, and the rectum

  - peritoneum is continuous with that of the abdominal cavity

H. The **four abdominopelvic quadrants** (more general method of localizing the visceral organs) delineate the abdominopelvic cavity into four segments by drawing one horizontal plane and one vertical plane through the umbilicus.

  o Right upper quadrant (RUQ)
  o Right lower quadrant (RLQ)
  o Left upper quadrant (LUQ)
  o Left lower quadrant (LLQ)

I. The **nine abdominopelvic regions** (used by clinicians to map the visceral organs) are created by two transverse planes and two parasagittal planes, forming a "tic-tac-toe" grid.

  o UMBILICAL region -- the center square
  o HYPOCHONDRIAC regions (right and left) -- superior lateral regions
  o EPIGASTRIC region -- medial and superior to the umbilical region
  o LUMBAR regions (right and left) -- middle lateral regions
  o HYPOGASTRIC region -- medial and inferior to the umbilical region
  o INGUINAL (ILIAC) regions (right and left) -- inferior lateral regions

ADDITIONAL ANATOMICAL TERMS:

- **Absorption**: the route through which substances (only very small molecules) can enter the body, dependent upon catabolic reactions

- **Adaptability**: long-term responsiveness

- **Adaptation**: the change in living organisms that allow them to live successfully in an environment

- **Differentiation**: the process by which a less specialized cell becomes a more specialized cell type

- **Excretion**: the process of removing metabolic waste products and other useless materials

- **Growth**: refers to an increase in some quantity over time, often due to an increase in the size and/or number of individual cells

- **Metabolism**: the set of chemical reactions that occur in living organisms in order to maintain life

    - **Anabolism** refers to the construction of molecules, via metabolic pathways, from smaller units.

    - **Catabolism** refers to the breakdown of molecules, via metabolic pathways, into smaller units, consequently releasing energy.

- **Reproduction**: the process through which new individual organisms are produced; therefore, it is essential to the continuity of life.

- **Supine**: The patient is lying down with the face up.

- **Prone**: The patient is lying down with the face down.

- **Responsiveness**: the ability of an organism to change activity or functioning, based upon the application of a stimulus; also referred to as <u>irritability</u>

Chapter 2 – THE CELL

INTRODUCTION

- All living organisms are composed of **cells**, the basic structural and functional units of life.

- **CELL THEORY**:
  - Cells are the basic unit of structure in all living things.
  - New cells are formed/produced from other pre-existing cells, via division.
  - Cells are the fundamental units of structure that perform all vital functions.

- Two main types of cells in the body:
  - SOMATIC CELLS – body cells
  - SEX CELLS – reproductive cells or germ cells

- Cellular Diversity – the trillions of cells in the human body are made up of 200 different cell types that vary greatly in size, shape and function.

- **Cytology**: the study of the cell's structure and function
  - **Light Microscopy (LM)** – uses light to magnify and view cellular structures up to 2000x their natural size
  - **Electron Microscopy (EM)** – uses electrons to magnify and view cell ultrastructures up to 2 million times their natural size

- The 3 main parts of the cell: 1) Plasmalemma (plasma or cell membrane), 2) cytoplasm, 3) nucleus

CELL MEMBRANE (PLASMA MEMBRANE or PLASMALEMMA)

- **STRUCTURE**

  - FLUID MOSAIC MODEL concept:
    o A thin layer of extracellular fluid surrounds a cell.
    o Its outer boundary is a selectively permeable lipid bilayer called the **cell membrane** (also called **plasma membrane**, **plasmalemma**, or "phospholipid bilayer").
    o This cell membrane model is a bilayer of lipid molecules with protein molecules dispersed within it.

  - **INTEGRAL PROTEINS**: embedded in the phospholipid bilayer

  - **PERIPHERAL PROTEINS**: attached to the membrane but can separate from it

  - **CHANNELS**: allow water and ions to move across the membrane
    - **Gated channels** can open and close.

- **MICROVILLI**: tiny fingerlike projections of the cell membrane that increase the surface area of cells, and are involved in a wide variety of functions, including absorption, secretion, and cellular adhesion.

- **FUNCTIONS**

  - **PROTECTION**: forms a barrier against substances and forces outside the cell

  - **STRUCTURAL SUPPORT**: physical interconnections between individual cells occur, as well as connections to their extracellular environment

  - **SENSITIVITY**: some membrane proteins act as **receptors**, a component of the cellular communication system

  - **REGULATION of EXCHANGE with the environment (i.e., selective permeability)**: free passage of some (not all) materials are permitted

    - **DIFFUSION**: net movement of <u>material</u> from an area of high concentration to an area of low concentration; occurs until equilibrium is achieved (i.e. concentration gradient is eliminated)

    - **OSMOSIS**: net movement (or diffusion) of a <u>solvent</u> (frequently water) across a semi-permeable membrane, from a solution of low solute concentration (high water potential) to a solution with high solute concentration (low water potential)

    - **FACILITATED DIFFUSION (or facilitated transport)**: a process of diffusion; a form of passive transport facilitated by the presence of <u>transport (or carrier) proteins</u>

    - <u>**Active membrane processes**</u> are the mediated processes of moving molecules and other substances across the cell membrane, often requiring energy in the form of **ATP** (*adenosine triphosphate*).

      - **ACTIVE TRANSPORT**: energy-dependent (require ATP) and independent of concentration gradients; some ion pumps are exchange pumps

      - **ENDOCYTOSIS**: a process where cells absorb material (molecules such as proteins) from the outside by engulfing it with their cell membrane.

        - **PHAGOCYTOSIS**: cell eating (the process by which cells ingest large objects, such as bacteria or viruses)

        - **RECEPTOR-MEDIATED ENDOCYTOSIS**: a more specific active event where the cytoplasm membrane folds inward to form coated pits

        - **PINOCYTOSIS**: cell drinking (uptake of solutes and single molecules, such as proteins)

THE CYTOPLASM consists of three major elements: cytoplasm, organelles, and inclusions.

- **CYTOSOL** – an intracellula fluid that contains dissolved solutes, and surrounds the cellular organelles

- **ORGANELLES** – specialized subunits within a cell that has specific functions

  - **MEMBRANOUS ORGANELLES**: separately enclosed within their own lipid membranes that isolate them from the cytosol

    a. **The NUCLEUS** – contains the cell's chromosomal DNA
      - Surrounded by the NUCLEAR ENVELOPE (double-layered membrane)

      - Contains the fluid NUCLEOPLASM

      - Contains NUCLEOLI (the site of ribosomal RNA synthesis; singular=*nucleolus*)

      - Contains CHROMATIN (chromosome in the *non-coiled* state when the cell is not dividing)

      - Functions as the **control center of the cell**

      - Its genetic material, **DNA, directs the cell's activities** by providing the instructions for protein synthesis

      Also responsible for **transmitting genetic information**

    b. **MITOCHONDRIUM** (plural = *mitochondria*) - bean-shaped organelles, which are described as **"cellular power plants"** because they generate most of the cell's supply of ATP (95% of the supply), used as a source of energy

      - Consists of two membranes:
        - The <u>outer mitochondrial membrane</u>, which encloses the entire organelle and

        - The <u>inner mitochondrial membrane</u> that folds inward to produce **CRISTAE**, which increase the surface area and **enhance the organelle's ability to produce ATP**

      - **Mitochondrial MATRIX** – the space enclosed by the inner membrane
        - **Important in the production of ATP** with the aid of the ATP synthase contained in the inner membrane

    c. **ENDOPLASMIC RETICULUM (ER)** – an interconnected network of tubules, vesicles and cisternae (rounded chambers), which is involved in synthesis, storage, transport, and detoxification

o **ROUGH ER (RER)** has attached ribosomes, where proteins are assembled and packaged in <u>transport vesicles</u> to be exported to the Golgi apparatus. The rough ER has several **functions in making all proteins that are secreted from cells**, on its ribosomes, and **providing RER membrane for the cell membrane.**

o **SMOOTH ER (SER)** does not have attached ribosomes. It functions in **lipid and carbohydrate (steroid) synthesis, lipid metabolism, calcium ion storage, and drug detoxification.**

d. **GOLGI APPARATUS** (also called the *Golgi body*, *Golgi complex*, or *dictyosome*) – composed of flattened membrane-bound stacks known as <u>cisternae</u> (singular = *cisterna*)

o **Functions in packaging materials for lysosomes, peroxisomes, secretory vesicles, and membrane segments that are used to replenish the cell membrane**

o Transport vesicles from the RER are processed from the *cis*-**Golgi (convex end)** to *trans*-**Golgi (concave end)** through the apparatus.

o Secretory products are discharged from the cell through the process of **exocytosis** (the ejection of cytoplasmic materials by fusion of a membranous vesicle with the cell membrane)

e. **LYSOSOMES** – spherical, membrane-walled sacs that **contain digestive enzymes (acid hydrolases)**

o **Function in digesting unwanted substances**, such as excess or worn-out organelles, food particles, and engulfed viruses or bacteria

f. **PEROXISOMES** – smaller, membrane-walled sacs that contain enzymes, especially <u>oxidases</u> and <u>catalases</u>, that **function in removing toxic *peroxides* from the body**

o *Peroxide*: a compound that contains an oxygen-oxygen single bond

• <u>**NON-MEMBRANOUS ORGANELLES**</u>: not enclosed within their own lipid membranes; hence, they are always exposed to the cytosol

a. **RIBOSOMES** – complexes of two subunits (ribosomal RNA and protein); function as the **site of protein synthesis**; two types found in cells

• **Free** ribosomes – located within the cytosol
• **Fixed** ribosomes – bound to the ER

b. **CYTOSKELETON** – an elaborate, internal network of protein rods spanning throughout the cytosol; confers **strength and flexibility** to the cytoplasm; provides **support and shape to the cell**, as well as **intracellular movement**

o **MICROTUBULES** – composed of the protein tubulin; function as the main support of the cell; allow the cell to change shape; allow organelle movement; function during cell division in moving and separating DNA strands

o **MICROFILAMENTS** – mainly composed of thin strands of the actin protein

o **INTERMEDIATE FILAMENTS** – function in providing strength, stabilization of the organelles, and transport of materials within the cytoplasm

o **THICK FILAMENTS** – mainly composed of the protein myosin; produce movement with the action of actin

c. **CENTRIOLES** – long, barrel-shaped microtubules that radiate from the centrosome (spherical structure in the cytoplasm near the nucleus) in non-dividing cells

o Active in dividing cells and **function in directing the movement of chromosomes during cell division**
o Also function in **forming the bases of cilia and flagella**

d. **CILIA** – microtubules containing tail-like projections that are anchored by a basal body
o **Function in movement of fluids or secretions across the cell surface**, by beating rhythmically

e. **FLAGELLA** – a whiplike projection that **functions in moving a cell through surrounding fluid**
o This is a longer version of the cilium; **found only on sperm**, enabling them to swim in the female reproductive tract.

f. **MICROVILLI** (sing. = *microvillus*) - tiny fingerlike projections of the cell membrane that increase the surface area of cells

- ⬛ **INCLUSIONS** - chemical substances in the cytoplasm that may or may not be present in a cell, depending on the cell type

  - These are often stored nutrients, secretory products, and pigment granules

- ⬛ **INTERCELLULAR ATTACHMENTS** :
There are three different types of cell junctions, by which cells attach to each other or to extracellular protein fibers.

  - **OCCLUDING JUNCTION** (or **TIGHT junction**) – the lipid portions of the plasma membranes bind together to seal the intercellular space between the cells, thereby preventing materials from passing between them

- **COMMUNICATING JUNCTION (or GAP junction)** – membrane or channel proteins, called connexons hold two cells together, forming a narrow passageway (channels) between the cells

- **ANCHORING JUNCTION** – two adjacent cells are mechanically linked together at their lateral or basal surfaces

    o **DESMOSOME** (also called *macula adherens* or *macula adherentes*) – a system of *cell adhesion molecules* (**CAMs** = large transmembrane proteins that bind cells to each other and to the extracellular fluid) and intercellular cement that glue adjacent cells together

    o **HEMIDESMOSOME** – a very small stud- or rivet-like structures on the inner basal surface that attaches a cell to the filaments and fibers of the extracellular matrix

---

THE CELL LIFE CYCLE

- **CELL DIVISION:** the series of events that take place in an eukaryotic (nucleated) cell, leading to its replication; divided into two brief stages

    o **INTERPHASE is divided into the G1, S, G2 Stages**
      - **G1:** cells are metabolically active, make proteins rapidly, and grow vigorously
      - **S:** DNA replicates itself
      - **G2:** final part of interphase where enzymes needed for cell division are synthesized

    o *M* **PHASE,** or *mitotic phase* (cells divide in this phase) composed of two <u>tightly-coupled processes</u>:
      - **MITOSIS** (or *mitosis proper*) and **CYTOKINESIS,** which together define the division of the mother cell into two daughter cells, genetically identical to each other and to their parent cell

      - The **STAGES OF MITOSIS** (*mitosis proper*) in chronological order are Prophase, Metaphase, Anaphase, and Telophase.

> *Mnemonic devices: "PMAT" or "Please Make A Taco" = Prophase, Metaphase, Anaphase, Telophase*
> Another mnemonic device for memorizing the stages of mitosis including *interphase* and *cytokinesis* is "**I Party More At The Club,**" or "**IPMATC.**"
> Another way of remembering it is: "**In Paris Many Artists Teach**" = Interphase, Prophase, Metaphase, Anaphase, and Telophase.

- **MAJOR EVENTS in the stages of <u>MITOSIS PROPER</u> ("*PMAT*") and <u>CYTOKINESIS</u>:**

- **PROPHASE**
    o The **chromatin threads coil and condense**, forming barlike chromosomes, each of which are comprised of two identical chromatin threads, now called <u>chromatids</u>.

- o Chromatids are held together by a small, buttonlike body called a <u>centromere</u> and <u>cohesin</u> (a protein complex).
- o **Nucleoli disappear** as chromosomes appear.
- o Cytoskeletal microtubules disassemble.
- o The **mitotic spindle** (a new assembly of microtubules) **forms** between the centriole pairs.
- o The **nuclear envelope fragments** and begins to be dispersed to the ER.

- • **METAPHASE**
  - o Chromosomes cluster at the middle of the cell
  - o **Centromeres precisely align at the equator (exact center) of the spindle, at the metaphase plate.**

> **Memory aid: "M" is for "Metaphase"* and chromosomes align in the **"middle"** of the dividing cell

- • **ANAPHASE:** shortest stage of mitosis; typically lasts only a few minutes
  - o Begins abruptly as the **chromatid pairs separate** (centromeres of the chromosomes split), and each chromatid now becomes a chromosome in its own right.
  - o The **V-shaped daughter chromosomes move toward opposite ends** of the cell.

> **Memory aid: The **V-shaped chromosomes** also look like the letter "A" without the horizontal line; hence, **"A"-shaped** chromosomes could be associated with the **"Anaphase"** stage*

- • **TELOPHASE**
  - o Begins as soon as chromosomal movement stops
  - o Nuclear membranes form and the nuclei enlarge as the chromosomes begin to uncoil.
  - o Nucleoli reappear and the nuclei resemble those of interphase cells.

> **Memory aid: This final phase is like prophase in reverse.*

- • **CYTOKINESIS** completes the division of the cell into two daughter cells
  - o Occurs as a **contractile ring** of peripheral microfilaments forms at the **cleavage furrow** and **squeezes the cells apart**
  - o Actually begins during late anaphase and continues through and beyond telophase

Chapter 3 - TISSUES

INTRODUCTION

- **Tissues** are groups of closely associated specialized cells, which are similar in structure, and perform related (and limited) functions.

- **Histology** is the study of tissues.

- **Four PRIMARY TISSUE TYPES:**
  - Epithelial tissue (epithelium)
  - Connective tissue
  - Muscle tissue
  - Nervous (neural) tissue

EPITHELIUM (singular) / EPITHELIA (plural)

- Epithelium is a sheet of cells that covers a body surface or lines a body cavity.

- **FUNCTIONS:**
  - Protection
  - Sensory reception
  - Secretion
  - Absorption
  - Ion transport

- **SPECIAL CHARACTERISTICS**

  1. **Cellularity** – epithelia is composed of abundant densely-packed cells and very little extracellular material

  2. **Specialized contacts** such as cell junctions

  3. **Polarity**: apical and basal regions exist and differ in structure and function

  4. **Support** by underlying connective tissues

  5. **Avascular** but **innervated**

  6. **Regeneration**: due to high mitotic rates and the presence of mesenchymal (stem) cells, epithelia can regenerate.

- **EPITHELIAL SURFACE FEATURES**

    o **APICAL** Surface Features:
    - **Microvilli** - increase epithelial surface area; may anchor sheets of mucus

    - **Stereocilia** – very long microvilli that cannot move

    - **Cilia** – move fluid, usually mucus

    o **LATERAL** Surface Features:
    - **Cell junctions** – desmosomes, tight junction, gap junction

    o **BASAL** Surface Features:
    - **Basal lamina** – a sheet of proteins, which acts as a filter and as a scaffolding on which regenerating epithelial cells grow

    - **Basement membrane** – formed by the basal lamina plus some underlying reticular fibers

- **CLASSIFICATION of EPITHELIA**

    o Based on number of cell layers:
    - *Simple* (one layer of cells)

    - *Stratified* (2 or more layers of cells) – the superficial or apical layer is used to classify the epithelial type

    - *Pseudostratified with cilia* – a simple epithelium that contains both short and tall cells; classification used mainly in one type (pseudostratified columnar)

    o Based on cell shape:
    - *Squamous* (flattened cytoplasm and nucleus)

    - *Cuboidal* (spherical nucleus)

    - *Columnar* (oval or elongated nucleus, usually located basally)

    - *Transitional epithelia* are a stratified epithelium that stretches and changes shape due to the expansion of their cells' lumens (open spaces)

- **EPITHELIAL TISSUES**

    o **Simple Squamous Epithelium** – lines alveoli of the lungs
    - Forms the endothelium of blood vessels and the mesothelium of the ventral body cavity
    - Molecules rapidly diffuse (passive) through the delicate and thin layer of flat cells of this epithelium

- o **Stratified Squamous Epithelium** – <u>non-keratinized</u> type forms the <u>moist linings of the</u> <u>esophagus, mouth, and vagina</u>
  - <u>Keratinized</u> type forms the epidermis of the skin
  - <u>Protects</u> underlying tissues in areas subjected to <u>abrasion</u>

- o **Simple Cuboidal Epithelium** - occurs in <u>kidney tubules</u> and in <u>ducts and secretory portions of</u> <u>small glands</u>
  - Functions in <u>secretion</u> and <u>absorption</u>

- o **Stratified Cuboidal Epithelium** – occurs in largest <u>ducts of sweat glands</u>, <u>mammary glands</u>, and <u>salivary glands</u>
  - Functions in <u>protection</u>

- o **Simple Columnar Epithelium** – <u>non-ciliated</u> type lines the stomach and <u>intestines, gallbladder,</u> and excretory ducts of some glands
  - <u>Ciliated</u> type lines small bronchi, uterine tubes and some regions of the uterus
  - Functions in <u>absorption</u>, <u>secretion</u>, and <u>ion transport</u>

- o **Stratified Columnar Epithelium** – rare existence in the body
  - Small amounts occur in the male urethra and in large ducts of some glands
  - Functions in <u>protection</u> and <u>secretion</u>

- o **Pseudostratified (Ciliated) Columnar Epithelium** – <u>non-ciliated</u> type lines the <u>sperm-carrying</u> <u>ducts and ducts of large glands</u>
  - <u>Ciliated</u> variety lines the trachea and most of the upper respiratory tract
  - Function in <u>secretion</u> (especially mucus) and <u>propulsion of mucus by ciliary action</u>

- o **Transitional Epithelium** – lines the ureters, bladder, and part of the <u>urethra</u>
  - <u>Stretches</u> readily and permits <u>distension</u> of urinary organ by contained urine

- • **GLANDS** – many epithelial cells make and secrete a product (aqueous fluid containing proteins usually); such cells constitute glands

- o <u>**EXOCRINE**</u> glands secrete their products onto body surfaces (skin) or into body cavities
  - Contain <u>ducts</u> that carry secreted products to epithelial surface

  - **Serous glands** produce a watery solution that usually contains enzymes.

  - **Mucous glands** produce viscous, sticky <u>mucus</u>.

  - **Mixed glands** produce both types of secretions.

  - **Unicellular glands (GOBLET CELLS)** are the individual secretory cells that occur in epithelia containing scattered gland cells.

  - **Multicellular glands** occur as aggregations of gland cells (glandular epithelia) that produce exocrine or endocrine secretions.

- Classified by their ducts (simple or compound) and by the structure of their secretory units (tubular, alveolar or acinar, or tubuloalveolar)

- ### MECHANISMS of SECRETION

  - **APOCRINE** secretion - a portion of the secreting cell's body is lost during secretion (e.g. lactiferous glands in the breast)

  - **HOLOCRINE** secretion - the entire cell disintegrates to secrete its substances (e.g. sebaceous glands)

  - **MEROCRINE (ECCRINE)** secretion - cells secrete their substances by exocytosis (e.g. mucous and serous glands)

  - **ENDOCRINE** glands are <u>ductless</u> and secrete product (usually hormones) directly into the bloodstream.

CONNECTIVE TISSUE – the most diverse and abundant tissue type

- <u>FOUR CLASSES</u>:
  - ***Connective tissue proper***
  - ***Cartilage***
  - ***Osseous (bone) tissue***
  - ***Blood***

- Common embryonic origin is **mesenchyme**.

- <u>STRUCTURE</u>: cells separated from one another by a large amount of extracellular matrix (ground substance + fibers + tissue fluid)

  - The structure of blood is an <u>exception</u> in this case.

CONNECTIVE TISSUE PROPER

- **FIBROBLAST** – the most abundant cell type in connective tissue proper

  - Produces both the fibers and the ground substance of the extracellular matrix

  - <u>Three types of fibers in connective tissues</u>

    - *Collagen* fibers – resist tension

    - *Reticular* fibers – provide structural support

    - *Elastic* fibers – enable recoil of stretched tissues

- **<u>LOOSE (AREOLAR / ADIPOSE / RETICULAR) CONNECTIVE TISSUE PROPER</u>:**

  o **AREOLAR**: underlies almost all epithelia and surrounds capillaries
    - Ground substance and collagen, reticular and elastic fibers in the matrix surround fibroblast cells, fat (adipose) cells and defense cells.

  o **ADIPOSE**: similar to areolar but contains more adipose cells
    - Functions in increased nutrient-storage
    - White fat is abundant in the hypodermis.
    - Brown fat occur in babies, for generating heat and for heating the blood.

  o **RETICULAR**: similar to areolar structure but the only fibers in the matrix are reticular fibers
    - Occurs in bone marrow, lymph nodes, and the spleen
    - Form networks of caverns that hold free blood cells

  o **<u>FUNCTIONS</u> of Loose AREOLAR Connective Tissue Proper:**

    - Support and bind other tissues with its fibers

    - Hold tissue fluid in its jellylike ground substance

    - Fight infection with its many blood-derived defense cells, such as macrophages, plasma cells, and neutrophils

    - Store nutrients in fat cells

- **<u>DENSE (*) CONNECTIVE (or FIBROUS Connective) TISSUE PROPER</u>:**

  - Contains exceptionally thick collagen fibers and resists tremendous pulling forces

  o ***IRREGULAR**: similar to areolar but collagen fibers are thicker and run in different directions, appearing wavy in cross sections

    - Functions in resisting strong tensions from different directions

    - Occurs in the dermis and in organ capsules

  o ***REGULAR**: all collagen fibers in the matrix run in the same direction, separated by rows of fibroblasts, parallel to the direction of the pull

    - Main component of ligaments, tendons, aponeuroses and fascia

- o **ELASTIC*: similar to areolar but increased bundles of elastic fibers

  - Located in walls of arteries, around the bronchial tubes and within certain ligaments (e.g. connects successive vertebrae)

CARTILAGE

- Avascular and not innervated

- Matrix consists of thin collagen fibrils, ground substance and increased tissue fluid

- *Chondrocytes* (mature cartilage cells) reside in *lacunae* (cavities)

- **HYALINE** Cartilage – amorphous but firm matrix; collagen fibers predominate

- **ELASTIC** Cartilage – more elastic fibers in the matrix

- **FIBROCARTILAGE** – thick collagen fibers predominate

BONE (or OSSEOUS) TISSUE

- Hard, calcified matrix consists of inorganic calcium salts and contains collagen fibers

- **Osteoblasts** secrete the collagen fibers and ground substance of the matrix

- Mature **osteocytes** inhabit small pits or cavities called *lacunae* (singular = *lacuna*)

- Very well vascularized

BLOOD – classified as atypical connective tissue because of its structure; yet, it is a connective tissue type because it originates from mesenchyme

- **Red (RBC)** and **white (WBC) blood cells** are surrounded by nonliving liquid matrix called **blood plasma.**

EPITHELIAL and CONNECTIVE TISSUES FORM COVERING and LINING MEMBRANES:

- Consist of an epithelium plus an underlying layer of connective tissue

- Cover broad surfaces in the body

- **Cutaneous membrane** (skin) – dry membrane that covers the outer surface of the body

- **Mucous membrane (or mucosa)** – moist membrane that lines hollow internal organs that open to the body exterior

- **Serous membrane (or serosa)** – slippery membrane that lines closed pleural, pericardial, peritoneal cavities

MUSCLE TISSUE

- Composed of muscle cells containing many **myofilaments** (*actin* and *myosin*)

- Specialized to contract and generate movement

- Scant extracellular matrix separates the muscle cells

  - **SKELETAL** muscle: <u>multinucleated</u> muscle cells have a cylindrical and striated appearance due to highly organized arrangement of myofilaments

  - **CARDIAC** muscle: branching cells have a striated appearance; one nucleus; presence of <u>intercalated discs</u> (special cellular junctions)

  - **SMOOTH** muscle: no visible striations; one centrally-located nucleus; spindle-shaped cell; usually functions in propelling substances through hollow organs

NERVOUS (NEURAL) TISSUE

- Main component of the nervous (neural) organs: the brain, spinal cord and nerves

  - **NEURONS** – highly specialized nerve cells with extensions and processes that allow electrical impulse transmission

  - **NEUROGLIA (supporting cells)** – <u>non-conducting</u> cells that nourish, insulate and protect the neurons

PRACTICE QUESTIONS 1:
The Cell and Tissues

1. What is the major function of mitochondria?
    a. produce ATP
    b. manufacture proteins
    c. package proteins
    d. cellular digestion
    e. cell transport

2. Trabeculae are found in:
    a. osteons.
    b. compact bone.
    c. spongy bone.
    d. perforating canals.
    e. B and C

3. The primary site of protein synthesis is the:
    a. smooth ER.
    b. lysosome.
    c. ribosome.
    d. rough ER.
    e. mitochondrion.

4. In anatomical description, a person is prone when:
    a. lying on his or her side in a fetal position.
    b. lying face down in the anatomical position.
    c. standing with hands facing medially in the anatomical position.
    d. lying face up in the anatomical position.
    e. standing upright in the anatomical position.

5. The phagocytic cells in the epidermis are:
    a. Langerhans' cells.
    b. keratinocytes.
    c. melanocytes.
    d. Merkel cells.
    e. All of the above.

6. In osseous tissue, which of the following are found in lacunae?
    a. osteoclasts
    b. capillaries

    c. nerves
    d. chondrocytes
    (e.) osteocytes

7. The least common type of cartilage is:
    (a.) elastic cartilage.
    b. osteocartilage.
    c. hyaline cartilage.
    d. fibrocartilage.
    e. All of the above.

8. Which of the following cell types is *not* found in the epidermis?
    a. keratinocyte. ✓
    (b.) fibroblasts. ⌐ – connective tissue proper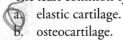
    c. melanocyte. ✓
    d. Merkel cell. ⌐
    e. Langerhans' cell. ⌐

9. Large, multi-nucleated fibers are found in:
    (a.) skeletal muscle.
    b. dense irregular connective tissue.
    c. cardiac muscle.
    d. smooth muscle.
    e. dense regular connective tissue

10. Which of the following pairs consists of anatomical opposites?
    a. proximal, medial
    b. cephalic, cranial
    c. distal, coronal
    d. medial, ventral
    (e.) dorsal, ventral

11. Membranes are organs formed by the combination of which tissues?
    a. epithelial and muscle
    b. muscle and neural
    c. muscle and epithelial
    d. connective and neural
    (e.) epithelial and connective

12. What is the term for molecule movement across a membrane from low concentration to high concentration?
    a. facilitated diffusion
    b. osmosis

diffusion = high → low

c. active transport
        d. filtration
        e. None of the above.

13. Recycling and changing the cell membrane is the major function of which organelle(s)?
        a. peroxisomes
        b. mitochondria
        c. nucleus
        d. Golgi (body) apparatus
        e. A and D

14. The two layers of a serous membrane are:
        a. pericardial and parietal.
        b. visceral and parietal.
        c. double sheets of peritoneum.
        d. pleural and parietal.
        e. A and B only

15. Which of the following is a part of the nucleus?
        a. DNA
        b. mitochondria
        c. histones
        d. plasmalemma
        e. A and C

16. A bone cell that produces new bone matrix is:
        a. stimulated by activity of the adrenal gland.
        b. an osteoclast.
        c. a chondrocyte.
        d. an osteocyte.
        e. an osteoblast.

17. Mature cartilage cells are termed:
        a. chondrocytes.
        b. chondroplasts.
        c. osteoblasts.
        d. osteoclasts.
        e. osteocytes.

18. Which of the following cells can be found in connective tissue?
        a. osteocytes
        b. macrophages
        c. Fibroblasts

d. All of the above.

e. A and B only

19. Which of the following statements is/are <u>true</u>?
    a. DNA forms nucleosomes with ribosomes.
    b. DNA controls the synthesis of RNA.
    c. DNA is bound to histones in the chromosomes.
    d. DNA replicates during the early stages of mitosis. *(interphase)*
    e. B and C

20. The wet membrane that <u>covers cavities</u>, which open to the external surface of the body is called the:
    a. cutaneous membrane. *Protection*
    b. synovial membrane.
    c. petechiae.
    d. mucous membrane.
    e. B and D.

21. Which of the following is/are <u>correct</u> regarding simple epithelia?
    a. They consist of a single cell layer. ✓
    b. They are avascular. ✓ *little or no blood vessels*
    c. They never have a free surface exposed to some inner chamber or passageway.
    d. They cover surfaces subjected to mechanical and chemical stress.
    e. A and B

22. The _____ lines the body wall within the abdominopelvic cavity.
    a. superficial fascia   d. parietal pericardium *-heart*
    b. mesocolon   e. parietal peritoneum
    c. mesentery proper

23. Which type of tissue is found in the capsules of organs?
    a. loose areolar connective tissue
    b. dense irregular connective tissue
    c. reticular connective tissue
    d. dense regular connective tissue
    e. adipose tissue

24. Serous cells undergo which mechanism of secretion?
    a. holocrine
    b. endocrine
    c. apocrine
    d. lactocrine
    e. merocrine

avascular - few or none
vascular - a lot blood vessels
innervated - nerves
not innervated - few or none

25. Which tissue is found in the walls of arteries and around bronchial tubes?
    a. dense irregular connective tissue
    b. loose areolar connective tissue
    c. elastic connective tissue
    d. dense regular connective tissue
    e. adipose tissue

26. Which tissue lines the ureters and part of the urethra?
    a. transitional epithelium
    b. dense regular connective tissue
    c. loose areolar connective tissue
    d. pseudostratified (ciliated) columnar epithelium
    e. simple squamous epithelium

27. Cytoskeletal microtubules disassemble during which phase of mitosis?
    a. anaphase
    b. metaphase
    c. telophase
    d. prophase
    e. interphase

28. Which cellular organelles are responsible for lipid metabolism?
    a. smooth endoplasmic reticulum (SER)
    b. lysosomes
    c. Golgi complexes (dictyosomes)
    d. peroxisomes
    e. rough endoplasmic reticulum (RER)

29. Which of the following does *not* occur during mitosis?
    a. Chromatids become chromosomes.
    b. Nuclear membranes form.
    c. DNA replicates itself. — interphase
    d. Chromatin threads coil and condense.
    e. None of the above. (All of the above occur during mitosis.)

30. Which intercellular attachment, also called "macula adherens," consists of a system of cell adhesion molecules (CAMs) and intercellular cement?
    a. occluding junctions
    b. anchoring junctions
    c. gap junctions
    d. desmosomes
    e. tight junctions

Chapter 4 – THE INTEGUMENTARY SYSTEM

A. INTRODUCTION

- **The integument (skin) and its appendages, or accessory structures, (hair, nails, sudoriferous or sweat glands, and sebaceous or oil glands) comprise the integumentary system,** which also consists of sensory receptors associated with nerve endings, allowing the sensations of touch, pressure, temperature and pain.

- The skin is the largest organ of the body, comprising approximately 7% of total body weight; measures 1.5 mm – 4 mm or more in thickness.

- The integument covers the entire surface of the body, including the eyes and eardrum.

- <u>All four primary tissue types are found in the integument.</u>

    - **An epithelium covers the surface.**

    - **Connective tissue provides underlying stability.**
        - Blood vessels are abundant within the connective tissue.

    - **Smooth muscle is found in the walls of blood vessels** in the integument, and attached to hairs.

    - **Nervous tissue controls the blood vessels as well as provides sensation.**

- FUNCTIONS of the integument includes:

    1. Physical protection
    2. Regulation of body temperature
    3. Excretion (and secretion)
    4. Nutrition (synthesis)
    5. Sensation
    6. Immune defense

- **There are 2 distinct layers of the skin: The Epidermis and The Dermis**

B. The **EPIDERMIS** : superficial layer = thick epithelium

KERATINIZED STRATIFIED SQUAMOUS EPITHELIA containing FOUR DISTINCT CELL TYPES:

- *Keratinocytes*: the most abundant cell type
  - They **produce keratin** (tough fibrous protein) that gives epidermis its protective properties.

  - <u>Arise from the deepest layer</u> of epidermis from cells undergoing almost <u>continuous mitosis</u>

  - Cells are <u>dead, flat sacs</u>, which are <u>completely filled with keratin</u> by the time they approach the skin surface.

  - In areas of friction, both cell production and keratin formation are accelerated.

  - Also <u>produce antibiotics and enzymes that detoxify</u> the harmful chemicals to which our skin is exposed

- *Melanocytes*: produce <u>melanin</u>, a dark skin pigment

- *Merkel cells*: intimately associated with sensory nerve ending fibers; may serve as <u>receptors for touch</u>

- *Langerhans cells*: macrophage-like <u>dendritic cell</u>
  - Function in policing our outer body surface, using receptor-mediated endocytosis to take up foreign proteins

LAYERS of the EPIDERMIS: listed in order from deep to superficial layers or *strata* (singular = *stratum*)

- THICK skin (5 strata); THIN skin (4 strata)

- *Stratum basale* or *Stratum germinativum* (*basal layer*):

  - Single row of cells consists of <u>basal cells (stem cell keratinocytes)</u>
  - Contains <u>Merkel cells</u> and <u>melanocytes</u>
  - **Continuous mitosis** occurs in this layer

- *Stratum spinosum* (*spiny layer*):

  - Comprised of keratinocytes, which contain thick bundles of <u>pre-keratin</u>
  - This layer also contains <u>Langerhans cells</u>.
  - Keratinocytes in this layer take on a spiky appearance due to the production of interconnecting proteins called **tonofibrils**, which increase stability in this layer.

- *Stratum granulosum* (*granular layer*):

  - 3-5 layers of flattened keratinocytes
  - Cytoplasm of keratinocytes contain:
    - **Keratohyaline** (help form keratin)
    - **Lamellated granules** (contain water-proofing glycolipid)

- *Stratum lucidum* (*clear layer* – does not stain well):

  - Transition zone that consists of flattened, densely packed cells, filled with keratin
  - *Found only in thick skin, which occurs only in the palms of the hands and soles of the feet*

- *Stratum corneum* (*horny layer*):

  - Consists mainly of **dead keratinocytes**, which are flat sacs completely filled with keratin
  - **Water-proofing glycolipids** exist in the extracellular space
  - Many cell layers thick (usually 15-30)
  - Exists as a thicker layer in thick skin
  - **Keratinized** is the term for an epithelium containing a large amount of keratin.
    - The relatively dry covering that results is relatively resistant to microorganism growth.
    - Keratinization occurs everywhere on the surface of the skin except the anterior surface of the eyes.

Mnemonic devices: "**Basil's** (or **German**) **Spiny Granules Look Corny**" --- to remember (*deep to superficial strata* in THICK skin)
Stratum **Bas**ale (or **Germ**inativum), S. **Spin**osum, S. **Granul**osum, S. **L**ucidum, S. **Corn**eum --- This phrase is actually most helpful since almost every word in the phrase gives an instant reminder to the term to remember, as indicated by the underscored letters.
Another phrase to remember the reverse order (*superficial to deep strata*): "**Could Lucid Grannies Spin Basketballs** (or **Gerbils**)?"

C. | The **DERMIS** | : the deep layer of connective tissue underlying the epidermis

- Strong, flexible connective tissue divided into two layers: *papillary* and *reticular layers*

- Cells found in this layer are *fibroblasts, macrophages, mast cells,* and scattered *white blood cells.*

- Fiber types are collagen, elastic and reticular

- Richly supplied with nerve fibers and blood vessels

- **Functions in nourishment and temperature regulation**

PAPILLARY LAYER: superficial portion (20%) of the dermis

- Comprised of **loose areolar** connective tissue proper

- **Dermal papillae** are finger-like pegs which project into the epidermis and lie atop dermal ridges.

- Folds in the *stratum germinativum* that extend into the dermis form **epidermal ridges**.

- Conversely, the dermis contains folds that fit in between the epidermal ridges called **dermal papillae**.

- **In thick skin**, on the palms and sole, the epidermal ridges and papillae are very pronounced and can be seen and felt on the surface, resulting in **fingerprints**, **palm prints**, and **footprints**.

- <u>Fingerprints and footprints increase the surface area</u> of the skin covering the digits, **increase friction and enhance the gripping ability of the hands and feet.**

- **Patterns** of epidermal ridges and dermal papillae are **genetically determined** and are different in everyone, and therefore can be used to identify individuals.

<u>**RETICULAR LAYER**</u>: deeper 80% layer of the dermis

- Comprised of **dense irregular** connective tissue

- Reticular due to the networks of **thick collagen fibers**, which extend into the papillary layer and deeper into the subcutaneous layer (*hypodermis*) to bind everything together.

- **Provides strength and resilience to skin**

- Blood vessels, glands, muscles, hair follicles and nerves are all found in this layer.

<u>**BLOOD SUPPLY of the INTEGUMENT**</u>

- **Cutaneous plexus** (plural = ***plexi***): network of blood vessels at the border of the reticular layer and the subcutaneous layer

- **Papillary plexus**: highly-branched network of blood vessels just deep to the epidermis

- These plexi play a role in <u>thermoregulation</u> and <u>overall blood flow</u>.

<u>**NERVE SUPPLY of the INTEGUMENT**</u>

- Nerve fibers <u>regulate blood flow</u> and <u>regulate the glands</u>.

- Provide **sensory perception**
  - <u>Tactile discs</u>: formed from the union of a Merkel cell and a sensory nerve ending

  - <u>Free dendrites</u> are sensitive to *pain* and *temperature*.

  - Other receptors of the skin
    - <u>Tactile corpuscles</u> (light touch)
    - <u>Root hair plexus</u> (light touch)

- ○ <u>Ruffini corpuscles</u> (stretch)
- ○ <u>Lamellated corpuscles</u> (deep pressure and vibrations)

D. | **SKIN COLOR** | is determined by a combination of three factors.

1. THICKNESS of the STRATUM CORNEUM

2. AMOUNT of PIGMENT(s) in the EPIDERMIS

- **<u>MELANIN</u>:** dark pigment (black, yellow-brown, or brown) produced and stored by the melanocytes

  - **MELANOSOMES** (vesicles containing melanin) are transferred intact from melanocytes to keratinocytes. This melanosome transfer colors the keratinocytes temporarily until the vesicles are destroyed by lysosomes.

  - **LIGHT SKIN** occurs if melanosomes are transferred in the stratum basale (germinativum) and stratum spinosum, and the cells of the more superficial layers lose their pigmentation.

  - **DARK SKIN** occurs if larger melanosomes are transferred in the stratum granulosum as well, causing darker and more persistent pigmentation.

  - Melanin helps to **prevent skin damage** by surrounding the nuclei of cells, so as to absorb ultraviolet (UV) radiation in sunlight.

  - Melanin production is triggered and increased by UV radiation and hence, leads to tanning.

- **<u>CAROTENE</u>: yellow-orange** pigment found in carrots and in green and orange leafy vegetables

  - May become trapped in the epidermis

  - Can be converted to <u>vitamin A</u>, which is required for epithelial maintenance and the synthesis of visual pigments in the eye

3. BLOOD SUPPLY in the DERMIS

- <u>Increased</u> blood flow causes "blushing."

- <u>Decreased</u> blood flow often causes the skin to become relatively pale.

- <u>Long-term decreased</u> blood flow causes **CYANOSIS** (a bluish coloration to the skin), resulting in **HYPOXIA** (low tissue O_2 concentrations) in the affected area.

E. | The **HYPODERMIS (subcutaneous layer or superficial fascia)** | : layer deep to the skin; <u>not part of the integumentary system</u> but shares some of the skin's functions

- Both areolar and adipose tissues occur in this layer; but, adipose tissue predominates.
- **Stores fat**
- **Anchors the skin to underlying structures** (mostly to muscles) but loosely enough so that skin can slide relatively free
- Acts as **insulator**
- Because no vital organs are in this region, it is a great location for injection of drugs; hence, the term *hypodermic needle*

F. | **APPENDAGES OF THE SKIN** |

HAIR and HAIR FOLLICLES:

- <u>**HAIR**</u>

  - The <u>main function of hair</u> is to **sense** things that lightly **touch** the skin, via the **root hair plexus.**

  - <u>Other functions:</u>
    - **Protection** of the scalp from UV light, blows to the head and **thermoregulation.**

    - **Trapping of particles** in the nasal passageway and external auditory canals.

    - <u>Arrector pili muscles</u> allow **movement of hairs,** as in "goose bumps or goose pimples."

  - Hard keratin predominates (soft keratin is found in typical epidermal cells)

  - The **hair root** (portion of the hair attached to the hair follicle) and **shaft** (portion exposed to the surface) are the main parts of hair.

  - Hair consists of <u>three concentric layers of keratinized cells</u>:
    - *Medulla* – central core containing large cells and air spaces; absent in fine hairs

    - *Cortex* – several layers of flattened cells

    - *Cuticle* – single outer layer of dead cells, which overlap like shingles; most heavily keratinized, for strength

  - <u>**HAIR FOLLICLES:**</u> organs that <u>form the hairs</u>

    - Extend from the epidermal surface into the dermis, forming a hair bulb in the deep end

- STRUCTURE:
  - <u>Internal root sheath</u>: layer of cells surrounding the hair root and deeper parts of the shaft.

  - <u>External root sheath</u>: located superficial to the internal root sheath and resembles the layer of the epidermis; and spans the entire length of the follicle

  - <u>Glassy membrane</u>: thickened basal lamina of the epidermis

- **ARRECTOR PILI MUSCLE** (associated with each hair follicle)

- The **hair PAPILLA** is an area of connective tissue at the base of the hair follicle.
  - The **hair BULB** is the area of epithelial cells that surround the papilla.

- **Hair MATRIX** is the specialized area of epithelium that grows the hair.

- **<u>TYPES of HAIR:</u>**

  - **LANUGO:** the first hairs produced before birth, of which most is shed before birth and is replaced by one of three types of adult hairs, vellus, intermediate, or terminal

  - **VELLUS** hairs: fine, colorless "peach fuzz" that covers most of the body

  - **INTERMEDIATE** hairs: thin, colored hairs commonly found on the appendages and groin

  - **TERMINAL** hairs: coarse, darkly pigmented hairs found on the head and face

- **<u>HAIR COLOR:</u>** determined by melanin production in the melanocytes found in the hair papilla

  - **Density of melanin determines hair color (black or brown).**

  - **Red hair** is the result of a **biochemically distinct form of melanin.**

  - Hormones and age play roles in hair color as well.

    - **Gray hair** results as pigment production decreases, usually a sign of aging.

    - **White hair** results from a lack of pigmentation <u>and</u> the presence of air bubbles within the medulla of the hair shaft.

SEBACEOUS (OIL) GLANDS

- The skin's oil glands occur over the entire body, except on the palms of the hands and the soles of the feet.

- Most are associated with hair follicles.

- Oily product is **sebum**, which functions in <u>lubricating hairs</u> and <u>preventing bacterial growth</u>.

- (Holocrine secretion: whole cells break up to form the product, released by exocytosis)

- <u>Sebaceous follicles</u> are large sebaceous glands that do not attach to a hair follicle.

SWEAT (*sudoriferous*) GLANDS

- Distributed over entire skin surface, except on the nipples and parts of the external genitalia

- <u>MYOEPITHELIAL CELLS</u>: small contractile cells that squeeze the secretion, or sweat, out of a sweat gland

- **ECCRINE (or MEROCRINE)** sweat glands: most numerous type, especially on the palms and soles

  - Small, coiled tubular glands

  - **Produce true sweat** (99% water and 1% electrolytes)

  - FUNCTIONS of SWEAT:
    - <u>Thermoregulation</u> – sweat cools the skin surface and reduces body temperature
    - <u>Excretion</u> of water and electrolytes
    - <u>Protection</u> via dilution of chemicals on the skin and via bacteriocidal effects

- **APOCRINE** sweat glands: mostly confined to the axillary, anal and genital areas

  - Larger than eccrine glands; large, coiled tubular glands that extend very deep into the dermis and even the hypodermis

  - Produce a special kind of sweat consisting of fatty substances and proteins, via merocrine secretion
    - Bacteria may act on the secretion, which changes its biochemical makeup, thereby, causing an undesirable odor

OTHER INTEGUMENTARY GLANDS

- **MAMMARY GLANDS**

  - Milk-producing glands in the breasts
  - Anatomically related to apocrine sweat glands
  - (*Mammary glands are discussed in greater detail in the Reproductive System section.*)

- **CERUMINOUS GLANDS**

  - Modified sweat glands in the external auditory canal
  - **CERUMEN (earwax):** formed from secretions of these glands and those of nearby sebaceous glands

NAILS

- Scale-like modification of the epidermis, which covers the dorsal tip of a digit (finger or toe)

- Corresponds to the superficial keratinized layers of skin

- **STRUCTURE:**
  - Nail <u>MATRIX</u> is the actively growing region.
  - Nail <u>BODY</u> is the actual nail.
  - Nail <u>BED</u> is the epithelium under the nail body.
  - Nail <u>ROOT</u> is a fold in the epidermis near the bone of the digit. The nail body grows from this region outward.
  - Nail <u>GROOVES</u> are the lateral borders of the nail body.
  - Nail <u>FOLDS</u> are the upward folds in the epidermis lateral to the nail grooves.
  - <u>EPONYCHIUM</u> is the cuticle, an extension of the proximal nail fold that covers the nail root.
  - <u>LUNULA</u> is the pale area near the eponychium.
  - <u>HYPONYCHIUM</u> is the fold of epidermis deep to the distal nail body.

Chapter 5 – THE SKELETAL SYSTEM I:
SKELETAL TISSUES and SKELETAL STRUCTURE

A. | CARTILAGES |

- **LOCATION**
  - External ear; cartilages of nose
  - Articular cartilages (cover ends of most bones at movable joints)
  - Costal cartilages (connect the ribs to the sternum)
  - Trachea; larynx, including epiglottis
  - Intervertebral discs
  - Pubic symphysis

- **STRUCTURE**

  - Firm, gel-like matrix contains complex polysaccharides called **chondroitin sulfates**, which form complexes with proteins, resulting in **proteoglycans**.

  - **CHONDROCYTES** (*mature cartilage cells*), which occupy cavities (or *pits*) called **LACUNAE** (*singular = lacuna*), are surrounded by an extracellular matrix of ground substance, fibers and water.

  - Physical properties of cartilage depend on the nature of the matrix, which vary in water content between 60-80% volume.

    - Collagen fibers provide <u>tensile strength</u>.

    - Combined characteristics of the extracellular fibers and the ground substance confer <u>flexibility</u> and <u>resilience</u>.

  - **Avascular** and **NOT innervated**

  - Cartilage tissue is surrounded by a fibrous **perichondrium** (outer *fibrous layer* of dense irregular connective tissue + inner *cellular layer*).

  - Perichondrium <u>resists outward expansion</u> of cartilage under pressure; also functions in the <u>growth and repair</u> of cartilage.

- **TYPES**

  1. **HYALINE CARTILAGE** (*collagen fibers predominate*):

     - Provides support through <u>flexibility</u> and <u>resilience</u>

     - Most cartilage in the body is hyaline.

2. **ELASTIC CARTILAGE** (*elastic fibers predominate*):

- <u>Tolerate repeated bending</u> (e.g. cartilage in external ear and in epiglottis)

3. **FIBROCARTILAGE** (*thick collagen fibers predominate*):

- Resists strong <u>compression</u> and strong <u>tension</u> forces

- | GROWTH MECHANISMS |

  - **APPOSITIONAL GROWTH** (*from outside*):

    - **Chondro<u>blasts</u>** (*cartilage-forming stem cells*) in perichondrium undergo repeated cycles of division to produce new cartilage tissue.

    - Stem cells of the inner layer differentiate into chondroblasts.

    - **Immature chondroblasts secrete matrix.**

    - As the matrix grows, more chondroblasts are incorporated (become mature chondrocytes) and are replaced by divisions of stem cells in the inner layer.

    - This mechanism **gradually increases the outer dimensions of the cartilage; hence, growth *from the outside*.**

  - **INTERSTITIAL GROWTH** (*from within*):

    - **Chondro<u>cytes</u>** within the cartilage divide, and their daughter cells secrete new matrix.

    - As daughter cells secrete additional matrix, they separate and consequently **expand the cartilage *from within*.**

  - **Neither growth mechanism occurs in adult cartilages.**

  - Most cartilages **cannot repair themselves** after a severe injury.

B. | BONES |

- | FUNCTIONS of BONES | (besides contributing to body shape)

  1. <u>Support</u> weight of the body and cradles its soft organs

2. Protection (e.g. The skull houses the brain; vertebrae -- spinal cord; rib cage -- thoracic organs.)

3. Movement: act as levers for muscles to pull on

4. Mineral storage: mainly calcium; also phosphate

5. Blood-cell formation:
   • Red bone marrow makes the blood cells.
   • Yellow bone marrow does not – it is a site for fat storage.

• CLASSIFICATION OF BONES and EXAMPLES OF EACH

(based on shape of the bone)

1. **LONG BONES**: elongated shape (not based on overall size); e.g. *femur* (thigh bone); *phalanges* and *metacarpals* (hand bones)

2. **SHORT BONES**: roughly cube-shaped; e.g. *carpals* (wrist bones); *patella* (also includes sesamoid bones)

3. **FLAT BONES**: thin, flattened, usually somewhat curved shape; e.g. most cranial bones of the skull; ribs; *sternum*; *scapula*

4. **IRREGULAR BONES**: of various shapes and do not fit into previous categories; e.g. *vertebrae*; *os coxae* (hip bones)

• GROSS ANATOMY OF BONES

1. **COMPACT BONE** (EXTERNAL layer) – smooth, homogenous appearance

2. **SPONGY BONE or Cancellous Bone** (INTERNAL) – honeycomb of **trabeculae** (singular = ***trabecula***: *a connective tissue partition that subdivides an organ*) with open spaces in between, which are filled with red and yellow bone marrow

3. **STRUCTURE OF TYPICAL LONG BONE**

   • **DIAPHYSIS**: shaft or long axis of the bone

   • **EPIPHYSES (singular = epiphysis)**:

     • The bone ends are called PROXIMAL EPIPHYSIS and DISTAL EPIPHYSIS.

     • Each articulating (joint) surface is covered with articular cartilage.

- **EPIPHYSEAL LINE:** <u>remnant of the epiphyseal plate</u>, which is a disc of hyaline cartilage that grows during childhood (*refer to cartilage growth in previous section*)

- **BLOOD VESSELS**

    - <u>NUTRIENT ARTERIES and VEINS</u> together run through NUTRIENT FORAMEN (hole in wall of diaphysis) to supply the diaphysis:

        - The arteries nourish the bone marrow and spongy bone, and their branches extend outward to help supply compact bone

    - <u>EPIPHYSEAL ARTERIES and VEINS</u> serve each epiphysis in the same way

- **MEDULLARY (or MARROW) CAVITY:** center of diaphysis

    - Contains no bone tissue, but filled with <u>yellow bone marrow</u>

- **SKELETAL MEMBRANES** (*internal* and *external* bone coverings):

    - **PERIOSTEUM:** membrane of connective tissue that **covers the entire outer surface of each bone** *except* the ends of epiphyses

        - Superficial layer of **dense irregular** connective tissue

        - Deep layer with **osteo<u>blasts</u>** (*bone-depositing cells*) and **osteo<u>clasts</u>** (*bone-destroying cells*)

        - RICHLY INNERVATED

        - VASCULARIZED (supplied by branches from nutrient and epiphyseal vessels)

        - **SHARPEY'S FIBERS:** thick bundles of collagen that secure the periosteum to the underlying bone

            - Also provide insertion points for tendons and ligaments

    - **ENDOSTEUM:** thinner membrane of connective tissue that **covers the trabeculae** of spongy bone (<u>internal bone surfaces</u>) and **lines the medullary cavity**

        - Also contains **osteoblasts and osteoclasts**

4. **STRUCTURE OF SHORT, IRREGULAR and FLAT BONES**

    - Same composition as long bones:

- Compact bone covered by periosteum

- Spongy bone covered by endosteum

- NO diaphysis

- NO medullary cavity

- Trabeculae of spongy bone is filled with bone marrow

- In flat bones: **DIPLOË** = internal spongy bone

5. **BONE DESIGN and STRESS**

- <u>Compression stress</u>: loading off center threatens to bend the bone by compressing the bone on one side and stretching it on the other side.

- Strong, compact bone tissue in external region resists maximal compression and tension forces.

- Compression and tension forces cancel each other out in internal spongy bone region.

- Spongy bone and marrow cavities serve to lighten the heavy skeleton and accommodate bone marrow.

- Trabeculae of spongy bone align along stress lines in organized pattern to provide support along the stress lines.

- ┌─────────────────────────────────────┐
  │ **MICROSCOPIC STRUCTURE OF BONE** │
  └─────────────────────────────────────┘

- <u>**COMPACT BONE structure:**</u>

  - **OSTEON (Haversian system):** basic functional unit

    - Long, cylindrical structure oriented parallel to the long axis of bones and to the main compression stresses

    - Also referred to as "weight-bearing pillars"

  - <u>General structure of osteon:</u> A central (Haversian) canal is surrounded by concentric *lamellae* (singular = *lamella*).

    - **LAMELLA:** concentric layer of bone matrix in which all the collagen fibers run in a single direction

- **CENTRAL (Haversian) CANAL:** lined by endosteum; vascularized; innervated

- **PERFORATING (Volkmann's) CANALS:** connect blood and nerve supply of periosteum to that of the central canals and the medullary cavity

- **OSTEOCYTES (osteoblasts surrounded by matrix):** spider-shaped mature bone cells which occupy *lacunae* (small cavities or chambers in bone matrix)

  - Their "spider legs" processes occupy **canaliculi** (thin tubes), which run through the matrix and connect lacunae to each other, and to the central canals

  - **Canaliculi** (singular = *canaliculus*): mode of communication and supplying osteocytes with nutrients

  - Osteocytes are <u>essential for maintaining the bone matrix.</u>

- **SPONGY BONE structure:**

  - Each trabecula contains several layers of **lamellae** and **osteocytes**, but **NO OSTEONS** or blood vessels of its own.

  - **Osteocytes** are <u>nourished by capillaries</u> located in the **endosteum** surrounding the **trabeculae.**

- **CHEMICAL COMPOSITION OF BONE TISSUE**

  - **BONE CELLS (osteocytes, osteoblasts and osteoclasts)**

  - Osteo**cytes** = mature bone cells that are completely surrounded by hard bone matrix; occupy lacunae

  - Osteo**blasts** = immature, bone-forming cells; synthesize osteoid via the process of <u>osteogenesis</u>

  - Osteo**clasts** = large, multinucleated cells that help dissolve the bony matrix through the process of <u>osteolysis</u>; regulate calcium and phosphate concentrations in body fluids

  - (<u>Osteoprogenitor cells</u> = mesenchymal cells that play a role in the repair of bone fractures)

  - **EXTRACELLULAR MATRIX of connective tissue, collagen fibers, ground substance, water and mineral crystals** – consists largely of crystals of *hydroxyapatite*

- **ORGANIC PART** is comprised of **collagen** fibers, which provides tensile strength

  - **OSTEOID**: the organic part of matrix BEFORE it mineralizes or calcifies

- **INORGANIC PART** is comprised of crystals of CALCIUM PHOSPHATE SALTS, which precipitate in the matrix and make the bone hard and able to resist compression

- | **BONE DEVELOPMENT** | (OSTEOGENESIS or OSSIFICATION)

- **OSSIFICATION:** the formation of bone tissue

- **CALCIFICATION:** the deposition of calcium salts within a tissue

1. **INTRAMEMBRANOUS OSSIFICATION** (also called *dermal ossification*) of embryonic mesenchyme

   - FORMS MEMBRANE BONES (which occur in FLAT BONES of the *skull* and the *clavicles*)

   - Begins when **osteoblasts differentiate** within a mesenchymal or fibrous connective tissue, at an **ossification center**

   - Ultimately produces spongy or compact bone

   - A network of bone tissue woven around capillaries (**woven bone tissue**) first appears and is then remodeled into a flat bone.

2. **ENDOCHONDRAL OSSIFICATION** of a hyaline cartilage model (starts in late embryonic period ~week 8)

   - This process FORMS ALL OTHER BONES = *endochondral* or *cartilage bones*.

   - Begins with the formation of a cartilaginous model

   - **Hyaline cartilage model is gradually replaced by osseous tissue.**

   - The <u>length</u> of a developing bone increases at the **epiphyseal cartilage** *(or epiphyseal plate)*, which separates the diaphysis from the epiphysis.

     - New cartilage is added at epiphyseal side.

     - Osseous tissue replaces older cartilage at diaphyseal side.

     - Time of closure of epiphyseal cartilage varies among bones and among individuals.

- Bone <u>diameter</u> grows via <u>appositional growth</u> at the outer surface of the bone.

- **ANATOMY of the EPIPHYSEAL GROWTH AREA**

  - The chondrocytes of the growing cartilage of the fetal epiphyses and the postnatal epiphyseal plates are organized into several zones, which allow rapid growth.

  - **Organization of the zones within the epiphyseal cartilage** (*epiphyseal* to *diaphyseal* side):

    - Zone of <u>R</u>esting cartilage
    - Zone of <u>P</u>roliferating cartilage
    - Zone of <u>H</u>ypertrophy
    - Zone of <u>C</u>alcification

Mnemonic device: "<u>R</u>ats <u>P</u>refer <u>H</u>ouses with <u>C</u>heese." – for the consecutive <u>zones</u> within the epiphyseal cartilage (refer to the list above)

- **POSTNATAL GROWTH of ENDOCHONDRAL BONES**

  1. Endochondral bones lengthen during growth through the growth of epiphyseal plate cartilages, which close in early adulthood.

  2. Bones increase in width through <u>appositional growth</u> (similar to appositional growth mechanism of cartilage).

     - Periosteum adds bone tissue to its surfaces, while the endosteum's osteoclasts remove bone from the internal surface of the diaphysis wall.

- **BONE REMODELING**

  - New bone tissue is continuously deposited and reabsorbed in response to *hormonal* (**parathyroid hormone** or **PTH**) or *mechanical stresses*.

  - **Mineral turnover and recycling** allow bone to adapt to new stresses

  - **Calcium** is the most common mineral in the body (>98% of it located in the skeleton), and is an important mineral for bone health.

  - Spongy bone in human skeleton is entirely replaced every 3-4 years; compact bone – every 10 years.

In the adult skeleton:

- **BONE DEPOSITION**: <u>**OSTEOBLASTS**</u> SECRETE OSTEOID on bone surfaces and CALCIUM PHOSPHATE SALTS CRYSTALLIZE within this osteoid

- **BONE REABSORPTION**: <u>**OSTEOCLASTS**</u> break down bone by SECRETING ACID (which dissolves the mineral part of the matrix) and LYSOSOMAL ENZYMES (which digest the organic part of the matrix)

    - This process **releases calcium and phosphate** into the blood.

    - **PTH** (*parathyroid hormone*) INCREASES REABSORPTION in response to decreased calcium levels in body fluids.

- COMPRESSION FORCES and GRAVITY acting on the skeleton HELP MAINTAIN BONE STRENGTH, as bones thicken at sites of stress.

- **REPAIR OF BONE FRACTURES** (cracks or breaks in a bone)

    - <u>SIMPLE</u> fracture does not penetrate the skin.

    - <u>COMPOUND</u> fracture penetrates the skin.

    - Treated by OPEN REDUCTION (surgical using wires and pins) or CLOSED REDUCTION (manual, using hands to realign bone ends). (reduction = realignment of broken bone ends)

    - Healing of a fracture can usually occur if portions of the blood supply, endosteum, and periosteum remain intact.

    - Healing stages:
        a. Fracture hematoma (large blood clot) formation
        b. Fibrocartilaginous callus formation (internal "soft callus")
        c. Bony callus formation
        d. Bone remodeling of the external callus into the original bone pattern

- **TYPES OF FRACTURES:**

    1. COLLES' FRACTURE: a break in the distal portion of the radius

    2. POTT'S FRACTURE: occurs at the ankle and affects both the tibia and the fibula

    3. COMMINUTED FRACTURE: shatters the affected area into a multitude of bony fragments

4. COMPRESSION FRACTURE: occur in vertebrae subjected to extreme stresses; more common when bones are weakened by osteoporosis

5. DEPRESSED FRACTURE: broken bone portion is pressed inward; typical of skull fracture

6. DISPLACED FRACTURE: a break that produces new and abnormal bone arrangements

7. NONDISPLACED FRACTURE: retains the normal alignment of the bones or fragments

8. EPIPHYSEAL FRACTURE: tends to occur where the bone matrix is undergoing calcification and chondrocytes are dying

9. GREENSTICK FRACTURE: only one side of the shaft is broken, and the other is bent; generally occurs in children whose long bones have yet to ossify fully

10. SPIRAL FRACTURE: produced by twisting stresses that spread along the length of the bone

11. TRANSVERSE FRACTURE: a break that occurs along the long axis of the affected bone

Chapter 6 - THE SKELETAL SYSTEM II: The AXIAL SKELETON

A. INTRODUCTION

- 206 bones of human skeleton form the axial and the appendicular skeletons

- The **AXIAL SKELETON (80 bones)** forms the long axis of the body; supports the head, neck and trunk; and protects the brain, spinal cord and thoracic organs

- Consists of 3 major regions:
  - The SKULL
  - The VERTEBRAL COLUMN
  - The BONY THORAX

B. THE SKULL

- Most complex bony structure of the skeleton, formed by cranial (cranium) and facial bones

- Mostly FLAT bones connected by SUTURES: All, but one, of the bones of the skull (the mandible) are joined by sutures.

- The **CRANIUM** (*"braincase"*) may be divided into two major areas:

  - CRANIAL VAULT (*calvarium*) – skull cap or roof of the skull

  - CRANIAL BASE or cranial FLOOR (inferior part) – **anterior, middle, and posterior cranial fossae** (a *depression*, singular = *fossa*)

- There are 85 named openings commonly called *foramina* (singular = *foramen*), *canals,* or *fissures,* and occasionally and *lacerum.*

- 22 bones comprise the skull: 8 cranial bones form the cranium and 14 facial bones comprise the rest

CRANIAL BONES (8)

1. **FRONTAL BONE (1):** "cockle-shell" shaped bone

   - Forms the forehead and the roofs of the orbits

   - GLABELLA: smooth part in the midline of frontal bone between the *superciliary arches,* which support the eyebrows

- FRONTAL SQUAMA: vertical anterior-most part of the frontal bone (i.e., *forehead*)

- <u>Supraorbital margins</u>: mark the superior limits of the orbits, the bony recesses that support and protect the eyeballs

- <u>Supraorbital foramen (notch)</u>: opening above each orbit, which transmits the supraorbital artery and nerve

- <u>Lacrimal fossa</u>: marks the location of the *lacrimal* (tear) *gland* that lubricates the surface of the eye

2. **PARIETAL BONES (2):** pair of curved rectangular bones

- Located posterolateral to the frontal bone, forming the sides of the cranium

- <u>Superior and inferior temporal lines</u>: low ridges on the external surface of each bone, marking the attachment of the temporalis muscle (closes the mouth)

- <u>Parietal eminence</u>: the smooth parietal surface superior to the temporal lines

- Four of the major sutures are associated with the parietal bones, at which they articulate with other cranial bones (*See SUTURES section*).

3. **OCCIPITAL BONE (1):** most posterior of the cranial bones

- Forms the floor and back wall of the skull

- FORAMEN MAGNUM: hole in the base of occipital bone through which spinal cord passes and is connected to the brain

- OCCIPITAL CONDYLES: facets on the base of the skull, which articulate with the superior facets of the C_1 (ATLAS) *vertebra* (plural = *vertebrae*)

- <u>Hypoglossal canals</u>: passageway for hypoglossal cranial nerve XII; begin at the lateral base of each occipital condyle

- <u>Basioccipital</u>: a band of bone anterior to the foramen magnum, which is the point of articulation between the occipital bone and the sphenoid

- <u>External occipital crest and protuberance</u>: midline prominences posterior to foramen magnum; the crest extends posteriorly from the foramen magnum, ending in the protruberance (a small midline bump)

- <u>Superior and inferior nuchal lines</u>: horizontal ridges that intersect the external occipital crest; these lines mark the attachment of muscles and ligaments that stabilize the articulation between the first vertebra, C_1, and the occipital condyles

4. **TEMPORAL BONES (2)**: inferior to parietal bones on lateral skull

Four Major Regions of the Temporal Bones: *Squamous / Tympanic / Mastoid / Petrous*

- **SQUAMOUS** region or part: the lateral surface bordering the squamous suture
  - SQUAMA: the convex external surface of the region

  - CEREBRAL SURFACE: the concave internal surface, whose curvature parallels the surface of the brain

  - ZYGOMATIC PROCESS: the inferior margin of the region, which curves laterally and anteriorly to meet the *temporal process of the zygomatic bone*; together, they form the **zygomatic arch** (or *cheekbone*), which defines the projection of the cheek

  - The squamous region abuts the parietal bone on each side.

- **TYMPANIC** region or part: surrounds the *external acoustic meatus* (or *external auditory canal*), immediately posterior and lateral to the mandibular fossa

  - Contains the STYLOID PROCESS, a needlelike projection inferior to the external auditory meatus

  - The **styloid process** marks the attachment site for ligaments that support the hyoid bone and for muscles of the tongue, pharynx, and larynx *(Some textbooks regard this landmark as a part of the petrous region instead.)*

- **MASTOID** region or part: "breast-shaped" area posterior to the ear *(Some textbooks do not regard this as a separate region; instead, it is incorporated with the petrous region.)*

  - Contains the MASTOID PROCESS, a rough projection, which anchors some neck muscles that rotate and extend the head

- **PETROUS** region or part: the most massive portion of the temporal bone -- contributes to the cranial base and forms the lateral region of the skull base

  - Contains the JUGULAR FORAMEN, passageway through which the internal jugular vein (largest vein of the head) and cranial nerves IX, X, and XI pass

  - CAROTID CANAL: opens in this region on the skull's inferior aspect, just anterior to the jugular foramen; passageway for internal carotid artery

  - FORAMEN LACERUM: jagged opening between the medial tip of the petrous portion of the temporal bone and the sphenoid bone; almost completely closed by cartilage in a living person, but conspicuous in a dried skull

  - INTERNAL ACOUSTIC MEATUS: lies in the cranial cavity on the posterior face of the petrous region; transmits cranial nerves VII and VIII

5. **SPHENOID (1)**: "bat-shaped" bone

- Forms the anterior plateau of the middle cranial fossa

- Spans the width of the cranial floor

- SPHENOID BODY + 3 PAIRS of PROCESSES:

  - GREATER WINGS – visible exteriorly anterior to temporal; form a portion of the orbits of the eyes

  - LESSER WINGS – bat-shaped portions located anterior to the sella turcica

  - PTERYGOID PROCESSES: vertical projections that begin at the boundary between the greater and lesser wings

    o Contain attachment sites for pterygoid muscles that move the lower jaw and soft palate

- <u>**SELLA TURCICA**</u> (resembles a "Turkish saddle") on the superior surface of the body

  - Contains the **hypophyseal fossa**, which holds the pituitary gland (**hypophysis**)

  - ANTERIOR CLINOID PROCESSES: located on either side of the sella turcica are these posterior projections of the lesser wings of the sphenoid

  - TUBERCULUM SELLAE: forms the anterior border of the sella turcica

  - DORSUM SELLAE: forms the posterior border

  - POSTERIOR CLINOID PROCESSES: extend laterally on either side of the dorsum sellae

  - The inferior surfaces of the sella turcica form part of the orbit and the superior part of the superior orbital fissure.

  - OPTIC GROOVE: transverse groove that crosses to the front of the saddle, above the seat

  - OPTIC CANAL: openings at either end of the optic groove

- <u>**FIVE IMPORTANT OPENINGS of the SPHENOID:**</u>

  - OPTIC FORAMEN – anterior to sella turcica
    o CN II passes through from the orbit into the cranial cavity

  - SUPERIOR ORBITAL FISSURE – long slit between the greater and lesser wings
    o Transmits CN III, IV and VI (*control eye movement*)

- FORAMEN ROTUNDUM – lateral to sella turcica
  - Passage for a branch of CN V

- FORAMEN OVALE – posterior to sella turcica
  - Passage for a branch of CN V

- FORAMEN SPINOSUM – inferior aspect
  - Transmits the middle meningeal artery, which supplies the broad inner surfaces of the parietal and temporal bones

6. **ETHMOID BONE (1):** irregularly shaped bone

- The most deeply situated bone located anterior to the sphenoid

- Forms most of medial bony area between the nasal cavity and the orbits

- Forms part of the orbital wall, the anteromedial floor of the cranium, the roof of the nasal cavity, and part of the nasal septum

- **Three major parts** (*cribriform plates, ethmoidal labyrinth*, and *perpendicular plate*):

  - CRIBRIFORM PLATES: contributes to roof of nasal cavities and floor of anterior cranial fossa

    - Contain OLFACTORY FORAMINA (or CRIBRIFORM FORAMINA), which transmit olfactory fibers of CN I (provide the sense of smell)

    - CRISTA GALLI: superior projection between the 2 cribriform plates; attached to *falx cerebri* (a membranous ligament), which helps secure the brain within the cranial cavity

  - ETHMOIDAL LABYRINTH: an interconnected network of ethmoidal air cells

    - Dominated by the SUPERIOR and MIDDLE NASAL CONCHAE, which are thin scrolls of bony structures that contribute to the conchae of nasal cavity, on either side of the perpendicular plate

    - Best viewed from the anterior and posterior surfaces of the ethmoid

  - PERPENDICULAR PLATE: forms superior part of NASAL SEPTUM

    - Projects inferiorly in the median plane

    - Superior portion is covered by olfactory epithelium

    - LATERAL MASSES: irregularly shaped, thin-walled bony regions flanking the perpendicular plate laterally; also riddled with sinuses

- **CRANIAL BONE SUTURES**

  - CORONAL SUTURE: where the parietal bones articulate with the frontal bone

  - SQUAMOUS SUTURE: where the parietal bones articulate with the temporal bones inferiorly

  - SAGITTAL SUTURE: where right and left parietal bones meet

  - LAMBDOID SUTURE: where the parietal bones articulate with occipital bone

  - FRONTONASAL SUTURE: boundary between the superior aspects of the two nasal bones and the frontal bone

  - OCCIPITOMASTOID SUTURE: where the occipital bone articulates with the temporal bones

FACIAL BONES (14)

1. **MANDIBLE (1):** lower jawbone

   - <u>Regions/Landmarks</u>:

     - MANDIBULAR BODY: horizontal portion that supports the teeth

     - RAMI of the MANDIBLE (singular = ramus): ascending portions from each side of the body

       o MANDIBULAR ANGLES: where each ramus meets the body

       o ALVEOLAR MARGIN or PART: thickened area that contains the alveoli and the roots of the teeth

       o CONDYLAR PROCESSES: posterior processes that enlarge superiorly to form the MANDIBULAR CONDYLES (or *heads* of the mandible), which articulate with mandibular fossae of the temporal bones to form the temporomandibular joint

       o CORONOID PROCESSES: jutting anterior portion of each ramus; site of *temporalis muscle* attachment

       o MANDIBULAR NOTCH: the depression that lies between the condylar and coronoid processes

- o MENTAL FORAMINA: prominent openings lateral to the midline, penetrating the body on each side; passageway for mental blood vessels and nerve that serve the lower jaw

- o MYLOHYOID LINE of mandible: lies on the medial aspect of the body, marking the origin of the *mylohyoid muscle*

- o SUBMANDIBULAR FOSSA: a depression inferior to the mylohyoid line, in which the *submandibular salivary gland* is located

- o MANDIBULAR FORAMINA: openings on the medial aspect of both rami; passageway for mandibular branch of CN V (for tooth sensation)

2. **MAXILLARY BONES or MAXILLAE (2)** – (singular = ***maxilla***):

- The largest facial bones that form the upper jaw and central part of facial skeleton

- Form part of the orbits also:
  - On each maxilla, the ORBITAL SURFACE provides protection for the eye and other orbital structures.

- **All facial bones, except the mandible, articulate with the maxillae.**

- PALATINE PROCESSES: form anterior part of the *hard palate* (bony roof of the mouth)

- FRONTAL PROCESSES: lateral aspect of the bridge of the nose; articulates with the frontal bone and the nasal bones

- ZYGOMATIC PROCESSES: articulate with zygomatic bones

- INFERIOR ORBITAL FISSURE: elongated opening within each orbit, formed by the maxillae and the sphenoid

- INFRAORBITAL FORAMEN: opening under each orbit, in the orbital rim; passageway for infraorbital nerves and vessels that serve the nasal region

- ALVEOLAR MARGINS or PROCESSES: the oral margins of the maxillae, which contain the upper teeth

3. **ZYGOMATIC BONES (2):**

- Lateral to the maxillae

- Form the cheekbones and part of the lateral rims of the orbits

- Contributes to the inferior orbital wall

- TEMPORAL PROCESS: articulates with *zygomatic process of the temporal bone to form the zygomatic arch*

- ZYGOMATICOFACIAL FORAMEN: located on the anterior surface of each zygomatic bone; transmits a sensory nerve innervating the cheek

4. NASAL BONES (2):

- Small, rectangular bones <u>forming the bridge of the nose</u>

- Articulates with the frontal bone at the *frontonasal suture*

5. LACRIMAL BONES (2):

- Delicate, finger-nail shaped bones

- Situated in the medial portion of each orbit, where it articulates with the frontal bone, maxilla, and ethmoid

- LACRIMAL GROOVE (or LACRIMAL SULCUS): a shallow depression that leads to a narrow passageway, called the ***nasolacrimal canal***

- NASOLACRIMAL CANAL - formed by the lacrimal bone and the maxilla

  - This canal encloses the <u>tear duct</u> as it passes toward the nasal cavity.

6. PALATINE BONES (2):

- Small, L-shaped bones

- HORIZONTAL PLATES: posterior part of hard palate; articulate with the maxillae

- PERPENDICULAR PLATES (vertical portion of the "L" shape of the bone): posterior part of the lateral walls of the nasal cavity and small part of the orbits

  - CONCHAL CREST: ridge on the medial surface, marking the articulation with the inferior nasal concha

  - ETHMOIDAL CREST: ridge on the medial surface, marking the articulation with the middle nasal concha of the ethmoid

- NASAL CREST: a ridge that forms where the right and left palatine bones interconnect, marking the articulation with the vomer

7. **VOMER (1):**

- Slender, plow-shaped bone in the medial plane of the nasal cavity

- Forms the posterior and inferior portions of the nasal septum

- Articulates with both the maxillae and palatine bones along the midline

8. **INFERIOR NASAL CONCHAE (2):**

- Thin, curved (*scroll-like*) bones protruding medially from the lateral walls of the nasal cavity

- Perform the same functions as the conchae of the ethmoid

> **SPECIAL PARTS of the SKULL to consider**

- **ORBITS** (or **ORBITAL COMPLEXES**) **of the eyes:**

  - Enclose and protect the eyes

  - Comprised of parts of 7 bones of the skull:

    o FRONTAL bone, MAXILLA, LACRIMAL bone, ETHMOID, SPHENOID, PALATINE bone, and ZYGOMATIC bone

- **NASAL COMPLEX** or **NASAL CAVITY:**

  - Constructed of bone and cartilage

  - Enclose the nasal cavities and include the *paranasal sinuses* (air spaces connected to the nasal cavities)

  - The frontal bone, sphenoid, ethmoid, vomer, maxillae, lacrimal bones, and the inferior nasal conchae contribute to the complex.

  - The bridge of the nose is supported by the maxillae and nasal bones.

- **PARANASAL SINUSES:**

  - Air-filled chambers that act as extensions of and open into the nasal cavities

  - Found in the maxilla, sphenoid, ethmoid and frontal bones

  - Serve to lighten skull bones, produce mucus, and resonate during sound production

- The **HYOID BONE**:

  - *Not a skull bone but considered in this section due to its location*

  - Located superior to the larynx and inferior to the skull

  - The only bone in the body that does not articulate directly with any bone.

  - Suspended by the **stylohyoid ligaments**

  - Its BODY and HORNS are attachment points for tongue and neck muscles that raise and lower the larynx during swallowing.

  - GREATER HORNS: larger process on the hyoid, which help support the larynx and serve as the base for muscles that move the tongue

  - LESSER HORNS: connected to the stylohyoid ligaments, from which the hyoid and larynx hang beneath the skull, like a swing from the limb of a tree

C. | **THE VERTEBRAL COLUMN** |

  - Extends from the skull to the pelvis, forming the body's major axial support

  - Surrounds and protects delicate spinal cord while allowing spinal nerves to issue from the cord via openings between adjacent vertebrae

  - Consists of 26 IRREGULAR BONES:

    - **7 CERVICAL** vertebrae ($C_1 - C_7$)

    - **12 THORACIC** vertebrae ($T_1 - T_{12}$)

    - **5 LUMBAR** vertebrae ($L_1 - L_5$)

    - **1 SACRAL** vertebra (or **SACRUM**) (5 fused bones, $S_1 - S_5$)

    - **1 COCCYGEAL** vertebra (or **COCCYX**) (4 fused bones, sometimes 3 or 5 bones)

  - The number of cervical vertebrae is constant in humans, the rest can vary in ~5% of the population.

> **Mnemonic device:** Use the common mealtime rule for remembering the number of cervical, thoracic and lumbar vertebrae, respectively --- *"Breakfast at __7__ am; Lunch at __12__ pm; and Dinner at __5__ pm"*
> The remaining vertebrae (sacral and coccygeal vertebrae) are single bones.

- **<u>INTERVERTEBRAL DISCS</u>** (shock absorbers):

  - The vertebrae are separated by pads of fibrocartilage that cushion the vertebrae and absorb shock.

  - These intervertebral discs are cushion-like pads composed of:

    - **Nucleus pulposus** (inner sphere): gelatinous and acts like a rubber ball

      - Absorbs compressive stress

    - **Annulus fibrosus**: outer collar of ~12 concentric rings of <u>ligament</u> (outer rings) and <u>fibrocartilage</u> (inner rings)

      - Contains the nucleus pulposus and limits its expansion when the spine is compressed

      - Also acts in binding the successive vertebrae together

- **<u>LIGAMENTS</u>** (connect bone to bone)

  - **Anterior longitudinal ligament**: prevents hyperextension

  - **Posterior longitudinal ligament**: prevents hyperflexion

  - **Ligamentum flavum**: elastic connective tissue, which stretches and recoils during flexion and extension of the body

- **GENERAL STRUCTURE of VERTEBRAE**

  - **BODY (*centrum*)**: anteriorly located rounded central portion of the vertebra

  - **VERTEBRAL ARCH**: a composite structure formed by two pedicles and two laminae (singular = ***lamina***)

    - **PEDICLES** ("little feet"): the sides of the arch, which are short bony walls that project posteriorly from the vertebral body

    - **LAMINAE**, singular = ***lamina*** ("sheets"): flat roof plates that complete the arch posteriorly

  - **VERTEBRAL FORAMEN**: vertebral body and arches form a central opening

    - Successive vertebral foramina **form the vertebral canal** through which the spinal cord is transmitted

- **SPINOUS PROCESS** (*vertebral spine*): median, posterior projection from the vertebral arch

- **TRANSVERSE PROCESS:** project laterally from each pedicle-lamina junction

- **SUPERIOR ARTICULAR PROCESSES:** form movable joints with **inferior** articular processes of vertebra located immediately <u>superior</u> to it

  - Typically <u>face toward</u> the spinous process

- **INFERIOR ARTICULAR PROCESSES:** form movable joints with **superior** articular processes of vertebra located immediately <u>inferior</u> to it

  - Typically <u>face away</u> from the spinous process

- **INTERVERTEBRAL FORAMINA:** lateral openings between adjacent vertebrae formed by notches on pedicles' superior and inferior borders

- REGIONAL VERTEBRAL CHARACTERISTICS

1. **CERVICAL VERTEBRAE: ($C_1 - C_7$)**

   - Small and wide side to side

   - Triangular vertebral foramen

   - Short, bifid spinous process projects directly posteriorly; increases in size from C_2 to C_7

   - <u>Transverse processes with transverse foramina</u>

   - <u>**ATLAS**</u> (**C_1 vertebra**):

     - Contains **no body** and **no spinous process**

     - Ring of bone consisting of anterior and posterior arches, with a <u>lateral mass</u> on each side with <u>articular facets</u> on its superior and inferior surfaces

     - Anterior and posterior <u>tubercles</u>

     - The superior articular facets receive the occipital condyles of the skull, thus "carry" the skull (named after Atlas in Greek mythology).

     - **Allows flexion and extension of the head on the neck** (*nodding "yes"*)

- <u>AXIS</u> (C_2 vertebra):

  - Contains a body and a spinous process

  - **Dens** (or **odontoid process**) projects superiorly from its body, fuses with the atlas during embryonic development, actually **contributes as "body" of the atlas**

  - Acts as a **pivot for the rotation of the atlas and the skull**; hence, it is named the axis

  - Allows the head to *shake side to side ("no")*

- There is **NO intervertebral disc** between the C_1 and the C_2 vertebrae.

2. **THORACIC VERTEBRAE:** $(T_1 - T_{12})$

   - LARGER and HEART-SHAPED overall

   - Bears two costal demifacets

   - CIRCULAR vertebral foramen

   - <u>Long, sharp spinous process projects inferiorly</u>

3. **LUMBAR VERTEBRAE:** $(L_1 - L_5)$

   - <u>Massive, kidney-shaped body</u>

   - TRIANGULAR vertebral foramen

   - <u>Short, blunt, rectangular spinal process projects directly posteriorly</u>

4. **SACRUM** (small of back)

   - Curved, triangular shape

   - Formed by 5 fused vertebrae $(S_1 - S_5)$

5. **COCCYX** (tailbone)

   - Small and triangular

   - Consists of 4 (3 or 5 sometimes) fused vertebrae

   - Provides slight support of pelvic organs, otherwise useless

D. | THE BONY THORAX | (Thoracic Cage)

- Protects heart, lungs and other thoracic organs

- Roughly cone-shaped

- Includes thoracic vertebrae posteriorly, <u>ribs</u> laterally, and <u>sternum</u> and costal cartilages anteriorly and medially

- | STERNUM | (breastbone) – anterior midline of thorax

    - **MANUBRIUM:** clavicular notches articulate with clavicles; also articulate with the 1st and 2nd ribs

    - **BODY:** 4 fused bones (fuse after puberty); notched on sides where the sternal body articulates with the 2nd to 7th ribs

    - **XIPHOID PROCESS:** inferior end of sternum

    - Exists as a plate of hyaline cartilage in youth, which does not fully ossify until approximately age 40

    - Process projects dorsally in some people

    - <u>3 important anatomical landmarks of the sternum:</u>

        o **JUGULAR NOTCH:** central indentation in superior border of manubrium

        o **STERNAL ANGLE:** horizontal ridge across anterior surface of sternum, where the sternal body joins the manubrium and forms a fibrocartilage hinge joint

            - Also used as a reference point for the 2nd rib and important landmark for thoracic surgery

            - 2nd intercostals space used for listening to certain heart valves

        o **XIPHISTERNAL JOINT:** where the sternal body and the xiphoid process fuse (T_9 level)

- | RIBS | (12 pairs form flaring sides of thoracic cage)

    - **TRUE RIBS (*Vertebrosternal* ribs):** superior 7 pairs (ribs 1-7), which attach directly to the sternum by their costal cartilages

- o   They progressively increase in length.

- **FALSE RIBS (*Vertebrochondral* ribs)**: inferior 5 pairs (ribs 8-12) <u>with INDIRECT</u> (ribs 8-10) <u>or NO ATTACHMENT</u> (ribs 11-12) <u>to the sternum</u>

- o   Ribs 8-12 progressively decrease in length

- **FLOATING RIBS (*Vertebral* ribs)**:   ribs 11-12 have <u>no anterior attachments to the sternum</u>

- o   Their costal cartilages lie embedded in muscles of the lateral body wall

- **COSTAL MARGIN**: inferior margin of the rib cage formed by costal cartilages of ribs 7-10

- <u>**TYPICAL RIB STRUCTURE**</u>:

  - Bowed flat bone

  - **1ˢᵗ rib is atypical:** flattened superior to inferior surfaces and quite broad

  - **HEAD** (wedge-shaped): articulates with vertebral bodies at two facets

  - **NECK**: short, constricted region just lateral to the head

  - **TUBERCLE**: knob-like projection just lateral to the neck on the posterior side

    - o   Articulates with the transverse process of the thoracic vertebra of the same number

  - **SHAFT**: the remainder of the rib laterally and medial end

  - **COSTAL GROOVE**: posterior, inferior groove, which serves as a passageway for nerves and vessels

Chapter 7 – THE SKELETAL SYSTEM III:
The APPENDICULAR SKELETON

PART 1 : BONES OF THE PECTORAL GIRDLE and THE UPPER LIMB or EXTREMITY
(The pectoral girdles attach the upper limbs to the trunk.)

A.   THE PECTORAL GIRDLE: clavicle(s) + scapula(e)

The paired pectoral girdle and its associated muscles form the shoulders.

- CLAVICLE (collarbone): slender, slightly curved long bone

  - STERNAL END ("ice cream cone" shaped, **medial end**) – attaches to manubrium

  - ACROMIAL END (flattened, **lateral end**) – articulates with scapula

  - TRAPEZOID LINE (visible on **inferior aspect**) – attachment site for ligament

  - CONOID TUBERCLE – attachment site for ligament

  - FUNCTIONS:
    o Provide attachment sites for muscles

    o Act as anterior braces or struts, which hold the scapulae and arms laterally away from the thorax

    o Transmit compression forces from upper limbs to the thorax

- SCAPULA(E) [shoulder blade(s)]: thin, triangular flat bones

  - BORDERS:

    o SUPERIOR border (shortest and sharpest)

    o MEDIAL border parallels the vertebral column

    o LATERAL border (thick) abuts the axilla and ends superiorly in a shallow fossa, the glenoid cavity

  - GLENOID CAVITY: shallow fossa or cavity which articulates with the head of the humerus

- o   Visible laterally and partially visible anteriorly

- • ANGLES of the SCAPULA:

   - o   SUPERIOR angle

   - o   LATERAL angle is thick and contains the glenoid cavity

   - o   INFERIOR angle moves as the arm is raised and lowered

- • SCAPULAR SPINE: bony ridge on posterior aspect of scapula

- • ACROMION: articulates with acromial end of clavicle

- • CORACOID PROCESS (bent-finger-shaped process): attachment point for *biceps brachii muscle* and *ligament attachment* to clavicle

- • SUPRASCAPULAR NOTCH: nerve passageway

- • FOSSA(E):   shallow depression(s)

   - o   INFRASPINOUS fossa on posterior aspect
   - o   SUPRASPINOUS fossa on posterior aspect
   - o   SUBSCAPULAR fossa formed by entire anterior surface of scapula

B. | **THE UPPER LIMB or EXTREMITY** (30 bones/limb) |

ARM or BRACHIUM (1 bone) / FOREARM or ANTEBRACHIUM (2 bones) / HAND (27 bones)

- • | **THE ARM or BRACHIUM** is comprised of the **HUMERUS** | (the largest and longest bone in the upper extremity).

   - • Articulates with the scapula at the shoulder and with the radius and ulna at the elbow

   - • <u>**PROXIMAL END FEATURES of the HUMERUS:**</u>

      - • HEAD (bulbous proximal end): fits into glenoid cavity of scapula

      - • ANATOMICAL NECK: constricted region located inferolateral to the head

- GREATER TUBERCLE: large projection on the lateral edge of the epiphysis of the humerus, forming the lateral margin of the shoulder; attachment site for *supraspinatus, infraspinatus, and teres minor muscles*

- LESSER TUBERCLE: lies on the anterior and medial surface of the epiphysis, marking the insertion point of the *subscapularis* muscle

- INTERTUBERCULAR GROOVE or SULCUS: separates the greater and lesser tubercles; guides a tendon of the *biceps brachii muscle* to its attachment point

- SURGICAL NECK: narrowed region located distal to the tubercles; most frequently fractured part of the humerus

- DELTOID TUBEROSITY: elevated surface that runs along the lateral border of the shaft, extending more than halfway down its length; attachment site for the *deltoid muscle*

- RADIAL GROOVE: runs along the posterior margin of the deltoid tuberosity; guides radial nerve of upper extremity

- **DISTAL END FEATURES of the HUMERUS:**

  - Articular CONDYLE: dominates the distal, inferior surface of the humerus; a low ridge that divides the condyle into two distinct articular regions – the trochlea and the capitulum

    - TROCHLEA (**medial** condyle): spool-shaped medial portion that <u>articulates with the ulna</u>

    - CAPITULUM (**lateral** condyle): rounded region that forms the lateral surface of the condyle, which <u>articulates with the radius</u>

  - MEDIAL and LATERAL EPICONDYLES: attachment sites for forearm muscles

  - MEDIAL and LATERAL SUPRACONDYLAR RIDGES

  - OLECRANON FOSSA (**posterior**): articulates with olecranon process of ulna

  - CORONOID FOSSA (**anterior**): accepts projections, along with olecranon fossa, from the surface of the ulna as the elbow approaches full flexion or full extension

  - RADIAL FOSSA (**anterior**): shallow depression superior to the capitulum; accommodates a small part of the radial head as the forearm approaches the humerus

- | **THE FOREARM or ANTEBRACHIUM: ULNA and RADIUS** |

  - Proximal ends of *radius* and *ulna* articulate with the *humerus*

- Distal ends articulate with the *carpus* (wrist)

- <u>The radius and ulna articulate with each other</u> both proximally and distally at the PROXIMAL and DISTAL RADIOULNAR JOINTS, respectively.

- INTEROSSEOUS MEMBRANE connects radius and ulna along their entire length

- SUPINATION: the radius (**lateral**) and the ulna (**medial**) are parallel

- PRONATION: the radius rotates medially over the ulna

- | ULNA | (forms the elbow):

  - Wide at the proximal end, then narrows at distal end; slightly longer than the radius

  - Main function is to <u>form the elbow joint with the humerus</u>

  - **PROXIMAL END FEATURES of the ULNA:**

    - OLECRANON and CORONOID PROCESSES (separated by the **trochlear notch**, a deep concavity) grip the TROCHLEA of the HUMERUS and form a hinge joint, allowing flexion and extension movements

    - RADIAL NOTCH: smooth depression where ulna articulates with head of radius

  - **DISTAL END FEATURES of the ULNA:**

    - HEAD of ulna – separated from the bones of the wrist by a disc of fibrocartilage; little or no role in hand movements

    - STYLOID PROCESS – attachment site for ligament to the wrist

- | RADIUS |

  - Thin at its proximal end and widened at its distal end – the opposite of the ulna

  - **PROXIMAL END FEATURES of the RADIUS:**

    - HEAD of radius: shaped like the end of a spool of thread

      - Articulates with *capitulum of humerus*

- • Medially articulates with *radial notch of ulna*, forming the PROXIMAL RADIOULNAR JOINT

  - • NECK of the radius: narrowed region that extends from the radial head to the radial tuberosity

- • RADIAL TUBEROSITY: attachment site for the *biceps brachii muscle*, which flexes (bends) the elbow, swinging the forearm toward the arm

- • **DISTAL END FEATURES**:

  - • ULNAR NOTCH (medial): articulates with head of the ulna, forming the DISTAL RADIOULNAR JOINT

  - • STYLOID PROCESS (lateral): anchors ligament to wrist

  - • Extreme distal end of radius is concave and articulates with carpal bones of the wrist.

  - • Radius contributes heavily to the wrist joint.

- • **THE HAND** (**27 bones**):

Carpus (8)/ Metacarpus (5)/ Digits (14)

- • **CARPUS** (true wrist): consists of eight marble-sized bones, called **CARPALS**, which are closely united by ligaments that are arranged in <u>two irregular rows of four bones each</u>:

  - • **PROXIMAL ROW:** LATERAL (thumb-side) to MEDIAL

    - • **SCAPHOID** – articulates with radius to form wrist joint

    - • **LUNATE** (*luna*, moon) – comma-shaped; articulates with radius to form wrist joint

    - • **TRIQUETRUM**, or **TRIQUETRAL** (triangular bone) – shaped like a small pyramid; articulates with the cartilage that separates the ulnar head from the wrist

    - • **PISIFORM** – smallest, pea-shaped bone that lies anterior to the triquetrum and extends farther medially than any other carpal bone in both the proximal and distal rows

- **DISTAL ROW:** LATERAL (thumb-side) to MEDIAL

  - **TRAPEZIUM** – lateral bone of the distal row; forms a proximal articulation with the scaphoid

  - **TRAPEZOID** – wedge-shaped; smallest distal carpal bone; also forms a proximal articulation with the scaphoid

  - **CAPITATE** - largest carpal; shaped like a head

  - **HAMATE** (*hamatum*, hooked) - contains a hook-like projection

---

Mnemonic Devices: Phrases to remember the proper order of the carpal bones, *proximal row* (*lateral to medial bones*) – *distal row* (*lateral to medial bones*), where the 1st letter of each word is the same 1st letter of each bone's name: *"**S**ally **L**eft **T**he **P**arty **T**o **T**ake **C**arlos **H**ome."* or *"**S**ome **L**overs **T**ry **P**ositions **T**hat **T**hey **C**annot **H**andle."* or *"**S**am **L**ikes **T**o **P**ush **T**he **T**oy **C**ar **H**ard."*

- CARPAL TUNNEL SYNDROME: The concave wrist bones are anteriorly covered by a ligamentous band that forms a tunnel through which the median nerve and muscle tendons run; INFLAMMATION of any element in the carpal tunnel (from overuse) compresses the median nerve, thereby causing pain or numbness.

- **METACARPUS** (palm)

5 metacarpals radiate distally from the wrist:

  - METACARPAL 1 (thumb-side) through METACARPAL 5 (little-finger-side)

  - The metacarpal bases articulate with the carpals proximally and with each other on their medial and lateral sides

  - Their bulbous heads articulate with the proximal phalanges of the fingers distally to form knuckles (METACARPOPHALANGEAL JOINTS)

- **DIGITS: PHALANX** (singular) or **PHALANGES** (plural)

14 phalanges (fingers/digits) per hand:

  - The thumb (POLLEX) has 2 phalanges: PROXIMAL PHALANX and DISTAL PHALANX

  - The rest of the fingers or digits have 3 phalanges each: PROXIMAL PHALANX, MIDDLE PHALANX and DISTAL PHALANX per digit

PART 2 : BONES OF THE PELVIC GIRDLE and the LOWER LIMB or EXTREMITY

A. **THE PELVIC GIRDLE** : attaches to axial skeleton by some of the strongest ligaments

- **COXAL BONES** ("hip bones" / also named *os coxae* or *innominate bones*): each of which consist of <u>three separate bones</u> during childhood, which <u>become fused in adults</u>

  - Boundaries of the three separate bones are indistinguishable but names are used to refer to the three different regions of the coxal bone

  - Y-shaped junction is formed where all 3 regions (**ilium**, **ischium** and **pubis**) meet

  - **ILIUM** : superior region of coxal bone

    - Consists of an inferior **BODY** and a superior, wing-like **ALA**

    - **ILIAC CREST** (thickened superior margin of ala): site of muscle attachment

    - **ILIAC SPINES** (4):

      o Posterior Superior Iliac Spine

      o Anterior Superior Iliac Spine

        ▪ <u>Prominent anatomical landmark</u>, which can be felt on the anterior of the hip.

      o Posterior Inferior Iliac Spine

      o Anterior Inferior Iliac Spine

    - **ACETABULUM:** located at Y-shaped junction of ilium, ischium and pubis

      ▪ Deep hemispherical socket that articulates with the ball-shaped head of the femur, forming the HIP JOINT

    - **GREATER SCIATIC NOTCH:** located posteriorly, just inferior to *Posterior Inferior Iliac Spine*

      ▪ A deep indentation through which the sciatic nerve passes, to enter the thigh

    - **ILIAC FOSSA:** concave internal surface of the iliac ALA

- **AURICULAR SURFACE:** roughened area (posterior to iliac fossa), which articulates with sacrum = SACROILIAC JOINT

- **ARCUATE LINE:** helps define the superior boundary of the true pelvis

- ISCHIUM : posteroinferior region (L-shaped or arc-shaped)

  - Consists of a thicker, superior **BODY** and thinner, inferior **RAMUS** (pl. = *rami*)

  - **ISCHIAL SPINE:** triangular projection, which is located posterior to the acetabulum and projects medially

    - Attachment point for SACROSPINOUS LIGAMENT (from the sacrum and coccyx)

  - **LESSER SCIATIC NOTCH:** inferior to ischial spine

    - Nerves and vessels that serve the perineum pass through this notch

  - **ISCHIAL TUBEROSITY:** rough, thickened area of inferior surface of ischial body

- PUBIS (paired V-shaped pubic bones): forms anterior region of coxal bone

  - **SUPERIOR PUBIC RAMUS:** branch of the pubis issuing from a flat body

  - **INFERIOR PUBIC RAMUS:** branch of the pubis issuing from a flat body

  - **PUBIC CREST:** thickened anterior border of pubic body

  - **PUBIC TUBERCLE:** knob-like lateral end of pubic crest

    - Attachment point for INGUINAL LIGAMENT

  - **OBTURATOR FORAMEN:** large opening between the pubis and the ischium

    - Few vessels and nerves pass through this opening

    - Foramen is almost completely closed by a fibrous obturator membrane

  - **PUBIC SYMPHYSIS** or **SYMPHYSIS PUBIS:** a fibrocartilaginous disc that joins the two pubic bones

  - **PUBIC ARCH:** formed by the inferior pubic rami and the ischial rami

- ▪ The angle of this arch helps to distinguish male from female *pelves* (singular = *pelvis*); e.g. wider angle in female *pelves*.

- ▪ Shallower and lighter female *pelves* provide more room in the true *pelvis*, for child-bearing purposes

B. | **THE LOWER LIMB or EXTREMITY** |

FEMUR (thigh) / CRUS (leg) / PES (foot):

- • | **FEMUR** | (THIGH bone)

The femur is the largest, longest, and strongest bone in the body. It can endure stress of 280 kilograms per square cm or 2 tons per square inch.

- • **PROXIMAL END FEATURES of the FEMUR:**

  - • **HEAD** of the femur: ball-like proximal end

  - • **FOVEA CAPITIS (*medial*):** small, central pit on femoral head

    - ▪ LIGAMENT of the head of the FEMUR runs from this pit TO THE ACETABULUM

  - • **NECK** of the femur: angles (~125°) laterally to join the shaft

    - ▪ The weakest part of femur

    - ▪ Often fractured in a "broken hip"

  - • **GREATER TROCHANTER (*lateral*):** projects laterally from the junction of the neck and shaft

    - ▪ Attachment site for various tendons and muscles

  - • **LESSER TROCHANTER (*posteromedial*):** originates on the posteromedial surface of the femur

    - ▪ Attachment site for various tendons and muscles

  - • **INTERTROCHANTERIC LINE (*anterior*):** interconnect the trochanters

- **INTERTROCHANTERIC CREST** (*posterior*): interconnect the trochanters

- **PECTINEAL LINE** (*medial*): inferior to the intertrochanteric crest

  ▪ Attachment site for the *pectineus muscle*

- **GLUTEAL TUBEROSITY** (*lateral*): on posterior surface of the shaft

  ▪ Attachment site of *gluteal muscle*

- **LINEA ASPERA** (*posterior*): a prominent elevation located on posteroinferior surface of the shaft

  ▪ Attachment site for the powerful hip muscles, the *adductor muscles*

- **FEMORAL <u>SHAFT</u>**: strong and massive, but curves along its length

  - Lateral bow facilitates weight bearing and balance

- **<u>DISTAL END FEATURES of the FEMUR</u>:**

  - **MEDIAL and LATERAL SUPRACONDYLAR RIDGE** (*posterior*): the linea aspera distally divides into these two ridges, to form a flattened triangular area, the POPLITEAL SURFACE

  - **LATERAL and MEDIAL CONDYLES** (*posterior* aspect): distal broadened area of the femur, <u>shaped like wagon wheels</u>

  - **LATERAL and MEDIAL EPICONDYLES:** the most raised points on the sides of the condyles

    ▪ Ligament attachment sites

  - **ADDUCTOR TUBERCLE:** bump on upper part of medial condyle

  - **INTERCONDYLAR FOSSA or NOTCH:** separates the two condyles posteriorly

  - **PATELLAR SURFACE:** separates the 2 condyles anteriorly

- **PATELLA** ("knee cap")

The patella is a triangular sesamoid bone enclosed within the tendon that secures the *quadriceps femoris* muscles of the anterior thigh to the tibia.

- FUNCTIONS:

  - Strengthens the *quadriceps tendon*
  - Protects the anterior surface of the knee joint
  - Increases the contraction force of the *quadriceps femoris*

- Rough, convex anterior surface

- Broad, superior **BASE** and a roughly pointed inferior **APEX**

- Posterior **FACETS** for medial and lateral condyles of the femur

- Posteroinferior **SURFACE for PATELLAR LIGAMENT**

- **CRUS (or LEG)**

The TIBIA and FIBULA comprise the bones of the leg.

- **TIBIA** : the second largest and strongest bone of the body after the femur

  - It receives the weight of the body from the femur and transmits it to the foot.

  - The <u>medial</u> bone of the leg

  - **FEATURES:**

    - **MEDIAL and LATERAL CONDYLES:** located at tibia's broad ***proximal*** end

      - They resemble two thick checkers lying side by side on top of the shaft.

      - Superior articular surfaces are slightly concave.

    - **INTERCONDYLAR EMINENCE:** an irregular projection that separates the two condyles

    - **TIBIAL TUBEROSITY** (***anterior*** aspect)**:** attachment site for PATELLAR LIGAMENT

    - **ANTERIOR BORDER or CREST:** sharp subcutaneous anterior ridge on the tibial shaft

- **MEDIAL MALLEOLUS** (plural = *malleoli*): inferior projection, which forms the MEDIAL BULGE of the ANKLE

  - Articulates with the TALUS bone of the foot

- **ARTICULAR SURFACE** (distal aspect): flat distal end of tibia, which articulates with TALUS of the foot also

- PROXIMAL TIBIOFIBULAR JOINT: in this joint, a **facet on inferior part of LATERAL TIBIAL CONDYLE** articulates with the FIBULA

- DISTAL TIBIOFIBULAR JOINT: in this joint, the **FIBULAR NOTCH on lateral side of distal tibia** articulates with the fibula

- $\boxed{\textbf{FIBULA}}$: thin, long bone with two expanded ends

  - Located <u>lateral</u> to the tibia

  - **FEATURES:**

    - **FIBULAR HEAD:** superior, proximal end

    - **LATERAL MALLEOLUS:** inferior projection which forms the LATERAL BULGE of the ANKLE

      - Articulates with TALUS of the foot

    - **SHAFT:** heavily <u>ridged</u>; appears to have been twisted a ¼ <u>turn</u>

    - <u>The fibula does not bear weight, but several muscles originate from it.</u>

- **Fractures most often occur at the medial and lateral malleoli of the tibia and fibula, respectively, because these are the least massive parts. These fractures are caused by inversion or eversion of the foot at the ankle.**

- $\boxed{\textbf{PES (or FOOT)}}$

The bones of the foot consist of the **TARSUS (ankle)**, the **METATARSUS (distal portion of the foot)**, and the **PHALANGES (digits or toes)**

- FUNCTIONS:

  - Supports the body's weight

  - Acts as a lever to propel body forward during walking and running

- SEGMENTATION makes the foot PLIABLE, making it adaptable to uneven ground

- The MEDIAL side of the foot is the HALLUX (**great toe**); opposite from the orientation of the hand where the POLLEX (thumb) is the lateral side, in the anatomical position.

- **TARSUS** (ankle): posterior ½ of the foot which contains 7 tarsal bones

  - **TALUS** ("ankle" part): the second largest bone in the foot

    - Transmits body weight from the tibia anteriorly, toward the toes

    - TROCHLEA of the talus: smooth superior surface, which contains lateral and medial extensions that articulate with the lateral malleolus (fibula) and medial malleolus (tibia), respectively

  - **CALCANEUS** ("heel bone" part): forms the heel of the foot

    - The largest of the tarsal bones

    - Carries talus on its superior surface

    - Thick tendon of calf muscles (CALCANEUS or ACHILLES TENDON) attaches to its posterior surface

    - CALCANEAL TUBEROSITY or TUBER CALCANEI: part of the calcaneus that touches the ground

    - SUSTENTACULUM TALI: medial, shelf-like projection, which articulates with the TALUS superiorly

  - **CUBOID** (*lateral*): "cube-shaped" tarsal

  - **NAVICULAR** (*medial*): "boat-like" tarsal

  - **MEDIAL / INTERMEDIATE / LATERAL CUNEIFORM BONES** (*anterior*):

    - "Wedge-shaped" tarsals arranged in a row

    - Located anterior to the navicular

    - Named according to their position

- METATARSUS : distal portion of the foot

  - **5 small, long METATARSALS in one foot**

  - Identified with Roman numerals - Metatarsals I-V (medial to lateral)

  - 1ˢᵗ metatarsal at base of big toe is the largest

  - Help support the body weight during walking, standing, and running

  - Forms the "ball" of the foot

- PHALANGES of the TOES: 14 phalanges in one foot

  - Smaller than those of the fingers; therefore, less nimble

  - General structure and arrangement are the same as those in fingers

  - HALLUX or DIGIT 1 (two phalanges): Distal Phalanx and Proximal Phalanx

  - DIGITS 2 - 5 (three phalanges each): Distal, Middle and Proximal Phalanges

Chapter 8 – THE SKELETAL SYSTEM IV: ARTICULATIONS (JOINTS)

> It is BEST to LEARN and KNOW the CLASSIFICATIONS of JOINTS BASED on FUNCTION and BASED on STRUCTURE; it is also advisable to LEARN the SUBTYPES of JOINTS and EXAMPLES of EACH.

A. **INTRODUCTION**

- Joints or articulations are *sites where elements of the skeleton meet and hold bones together* and allow *various degrees of movement.*

- Even though joints are the weakest part of the skeleton, there are factors that stabilize them:

    - Shapes of articulating surfaces

    - Ligaments

    - Tone of muscles whose tendons cross the joint

B. **CLASSIFICATION BASED ON FUNCTION**

(ARTHROSIS = ARTHROTIC JOINT; plural = ***arthroses***)

- **SYNARTHROTIC JOINT** (*synarthrosis*): allow no movement

- **AMPHIARTHROTIC JOINT** (*amphiarthrosis*): slight or limited movement

- **DIARTHROTIC JOINT** (*diarthrosis*): free movement

C. **CLASSIFICATION BASED ON STRUCTURE**
(*Fibrous / Cartilaginous / Bony Fusion / Synovial Joints*)

- **FIBROUS JOINTS**

    - Bones are connected by fibrous dense regular connective tissue

    - No joint cavity exists

    - Nearly all fibrous joints are <u>synarthroses</u> *(singular = synarthrosis)*

- ### 3 subtypes of Fibrous Joints

  - **SUTURE** (short fibers): *synarthrosis*; e.g. between flat bones of the skull

    - The edges of the bones interlock as they form this type of joint.

    - They may become completely fused later in life.

  - **SYNDESMOSIS** (long fibers):

    - *amphiarthrosis*; e.g. interosseous membrane between the radius and ulna

    - *synarthrosis*; e.g. anterior tibiofibular joint

  - **GOMPHOSIS** ("peg in socket"): *synarthrosis*; e.g. periodontal ligaments (tooth in socket)

- ### CARTILAGINOUS JOINTS

  - Bones are connected by cartilage

  - No joint cavity exists

  - Most cartilaginous joints are synarthroses and amphiarthroses

  - ### 2 subtypes of Cartilaginous Joints

    - **SYNCHONDROSIS** (hyaline cartilage): *synarthrosis*; e.g. epiphyseal plates, first rib-to-sternum

    - **SYMPHYSIS** (fibrocartilage): *amphiarthrosis*; e.g. intervertebral discs, pubic symphysis

- ### BONY FUSION

  - Two separate bones that have fused together to form a solid mass of bone

  - A totally rigid, immovable joint; therefore, a synarthrotic joint

  - No joint cavity exists

  - ### 1 subtype of Bony Fusion

    - **SYNOSTOSIS**: e.g. portions of the skull, such as along the frontal suture; epiphyseal lines

- **SYNOVIAL JOINTS**

  - Fluid-containing joint cavity exists

  - Joint is covered by articular cartilage

  - ALL SYNOVIAL JOINTS ARE **DIARTHROTIC.**

  - Most joints of the body, esp. those in the limbs, are in this class

  - **SIMPLE** synovial joint: most are of this subtype and contain two articulating surfaces

  - **COMPOUND** synovial joint: contain more than two articulating surfaces; (e.g. elbow, knee)

- **GENERAL STRUCTURE OF SYNOVIAL JOINTS**

  1. **ARTICULAR CARTILAGE** (*hyaline* cartilage): the ends of the opposing bones are covered by articular cartilage

  2. **JOINT (SYNOVIAL) CAVITY**: potential space that holds synovial fluid

  3. **ARTICULAR CAPSULE**: two-layered capsule enclosing joint cavity

     - Outer layer of **FIBROUS CAPSULE** of dense irregular connective tissue, which is continuous with the periosteum of bone

     - Inner layer of **SYNOVIAL MEMBRANE** (loose connective tissue) which lines the joint cavity

       o Covers all internal joint surface not covered by articular cartilage

       o Functions in PRODUCING <u>SYNOVIAL FLUID</u>

  4. **SYNOVIAL FLUID**: viscous, filtrate of blood arising from capillaries in the synovial membrane

     - Functions to ease movement at the joint

     - FLUID also OCCURS within the ARTICULAR CARTILAGES

5. **REINFORCING LIGAMENTS**: band-like ligaments (intrinsic, or capsular) that form the thickened parts of the fibrous capsule

- <u>EXTRACAPSULAR</u> ligaments are located just outside the capsule

- <u>INTRACAPSULAR</u> ligaments are located internal to the capsule

6. **NERVES and VESSELS**:

- Most nerves monitor joint stretching (stretch receptors).

- There are some pain receptors.

- Extensive capillary beds in synovial membrane produce blood filtrate, which is the basis of synovial fluid.

7. **ARTICULAR DISC**:

- *Intra-articular disc*, or *meniscus*, or **disc of fibrocartilage** in certain synovial joints that extends internally from the capsule and completely or partially divides the joint cavity in two

- OCCURS IN JOINTS WHOSE ARTICULATING BONES HAVE SOMEWHAT DIFFERENT SHAPES

- Functions in filling the gaps and improves the fit, thereby distributing the loading forces more evenly, minimizing wear and damage

- An example can be found in the knee joint = *meniscus (plural = menisci)*

- | **SYNOVIAL JOINT FUNCTION**: *decrease friction* |

  - **WEEPING LUBRICATION**: mechanism in which the cartilage-covered bone ends glide on a slippery film of synovial fluid squeezed out of the articular cartilages

- | **BURSAE and TENDON SHEATHS** |

  - THESE ARE NOT SYNOVIAL JOINTS; but they contain synovial fluid and are often associated with synovial joints (e.g. found in the shoulder joint)

  - **BURSA** (plural = *bursae*): flattened fibrous sac lined by a synovial membrane

    o Occurs where ligaments, muscles, skin, tendons or bones overlie each other and rub together

- **TENDON SHEATH**: elongated bursa that wraps around a tendon

  o Occurs only on tendons subjected to friction

- **SYNOVIAL JOINT MOVEMENTS**

Contracting muscles produce bone movements at synovial joints:

- **GLIDING** of one bone surface across another
  (e.g. *carpals*, *tarsals*, flat articular processes of *vertebrae*)

- ANGULAR MOVEMENTS: increase or decrease the angle between two bones

  o **FLEXION** (decrease angle)

  o **EXTENSION** (increase angle)

  o **HYPEREXTENSION** (bending beyond straight position)

  o *__AB__DUCTION** (movement away from the body midline)

  o **ADDUCTION** (movement toward the body midline)

  o **CIRCUMDUCTION** ("moving in a circle")

  o **ROTATION** (medial or lateral): turning movement of a bone around its own long axis

> ***Memory aid: Spell out the A-B in "ABduction" as "A-B-duction"** when reciting or recalling the word; and relate it to the word "<u>abduct</u>" or "<u>kidnap</u>" meaning "to take someone/something <u>away</u>." In this case, the meaning is "movement <u>away</u> from" the body midline.

- **SPECIAL MOVEMENTS OF SYNOVIAL JOINTS** (occur only at a few joints)

  - ***SUPINATION (turning laterally/backward)** vs. **PRONATION (turning medially/ forward)**: refer to movements of the radius around the ulna at the proximal radioulnar joint

  - **DORSIFLEXION** vs. **PLANTAR FLEXION**: up and down movements of the foot at the ankle

  - **INVERSION** vs. **EVERSION**: special movements of the foot

- **PROTRACTION** vs. **RETRACTION**: non-angular movements in the anterior and posterior directions

- **ELEVATION** vs. **DEPRESSION**: lifting superiorly vs. moving elevated part inferiorly

- **OPPOSITION**: unique action of the saddle joint of the thumb that allows grasping and manipulation of objects

> ***Memory aid:** To remember the movement "**supination**," think of cradling a bowl of **soup** (or a bowl of tea, as in "tea ceremonies") with one or both hands, to drink it. In this movement, the forearm (radius and ulna) is supinated, in which the **radius and ulna are parallel** to each other, as in the anatomical position. The opposite movement, "**pronation**," **moves the radius over the ulna in an X position**. This movement is achieved by actions such as placing your palms on the table or dribbling a basketball.

- **SYNOVIAL JOINTS CLASSIFIED BY SHAPE**

6 subtypes:

- **PLANE JOINT** (nonaxial): e.g. *intercarpal* and *intertarsal* joints, and joints between articular processes of vertebrae

  - Articular surfaces are essentially flat planes

  - Only short gliding movements allowed

- **HINGE JOINT** (uniaxial): e.g. elbow; ankle; joints between *phalanges* of fingers

  - Cylindrical end of 1 bone fits into a trough-shaped surface on another bone

  - Angular movement ("door on hinge")

- **PIVOT JOINT** (uniaxial): e.g. *proximal radioulnar* joint; articulation between C_1 and C_2 *vertebrae*

  - Rounded end of one bone fits into a ring formed by another bone plus an encircling ligament

  - Rotating bone can turn only around its long axis

- **CONDYLOID JOINT** (biaxial): e.g. wrist and knuckle (*metacarpophalangeal*) joints

  ○ Egg-shaped articular surface of one bone fits into an oval concavity in another

  ○ Allows moving bone to travel side by side, back and forth, but the bone cannot rotate around its own long axis

- **SADDLE JOINT** (biaxial): e.g. first *carpometacarpal* joint in ball of thumb

  ○ Each articular surface has both convex and concave areas, like a saddle

  ○ Allows the same movements as the condyloid joint does

- **BALL-AND-SOCKET JOINT** (multiaxial): e.g. shoulder and hip joints

  ○ Spherical head of one bone fits into a round socket in another

  ○ Allows movement in all axes, including rotation

- **TYPES of MOVEMENT DEPENDING on the CONSTRUCTION of the JOINT**

  ○ **UNIAXIAL** – movement in one axis or plane

  ○ **BIAXIAL** – movement in two axes or planes

  ○ **MULTIAXIAL** – movement in more than two axes or planes

  ○ **TRANSLATIONAL or NONAXIAL** – short gliding movements only

D. **SPECIFIC EXAMPLE: THE KNEE JOINT**

Please note that most <u>textbooks</u> discuss other representative articulations. For brevity and because it is the most complex articulation, only the knee joint is considered in this <u>study guide</u>.

- The largest and most complex joint in the body – it is <u>a complex of many types of joints</u>.

  ○ **COMPOUND and BICONDYLOID JOINT** because both the femur and tibia have two condylar surfaces

- o **PRIMARILY** acts as a **HINGE JOINT**, but allows **SOME MEDIAL and LATERAL ROTATION** when flexed and during leg extension

- o **FEMOROPATELLAR JOINT**: articulation between the patella and the distal/inferior end of the femur

- o **PLANE JOINT**, which allows patella to glide across the distal femur as the knee bends

- **SYNOVIAL CAVITY** has a complex shape with several incomplete subdivisions and several extensions leading to "blind alleys"

- More than a dozen bursae are associated with this joint.

  - Some examples are the **SUBCUTANEOUS PREPATELLAR BURSA, SUPRAPATELLAR BURSA; DEEP INFRAPATELLAR BURSA**

- <u>C-shaped menisci</u> (articular discs of fibrocartilage) occur within the synovial cavity:

  - **MEDIAL MENISCUS** and **LATERAL MENISCUS attach externally** to the **tibial condyles.**

  - **Both menisci have the following functions**:

    1. To facilitate uniform distribution of synovial fluid and of compression stress

    2. Stabilize the joint by guiding the condyles during flexion, extension, and rotation movements

    3. Prevent side-to-side rocking of the femur on the tibia

- **ARTICULAR CAPSULE** encloses the synovial cavity and can be seen on the posterior and lateral aspects of the knee

  - This capsule is absent anteriorly.

- Anterior aspect of the knee joint is covered by 3 broad ligaments, which run inferiorly from the patella to the tibia:

  - **PATELLAR LIGAMENT,** flanked by the **MEDIAL PATELLAR RETINACULUM** and **LATERAL PATELLAR RETINACULUM**

- Tendons of many muscles <u>reinforce the joint capsule and act as critical stabilizers of the knee joint</u>: **tendons of the *quadriceps femoris* and *semimembranosus*** muscles are the most important of these.

- The knee joint capsule is **further reinforced by several capsular and extracapsular ligaments**, all of which become taut when the knee is extended to prevent hyperextension of the leg at the knee.

- Two extracapsular ligaments are located on the lateral and medial sides of the joint capsule, the **FIBULAR and TIBIAL COLLATERAL LIGAMENTS**, respectively.

  - FIBULAR collateral ligament descends from lateral epicondyle of the femur to the head of the fibula.

  - TIBIAL collateral ligament descends from the medial epicondyle of the femur to the medial condyle of the tibia.

  - **Both halt leg extension and prevent hyperextension; also prevent lateral and medial movement of the leg at the knee.**

- Two strong intracapsular ligaments called ***cruciate ligaments***, which cross each other and run from the tibia to the femur, further stabilize the knee joint.

  - The **ANTERIOR CRUCIATE LIGAMENT (ACL)** arises from the anterior intercondylar area of the tibia and passes posteriorly to attach to the medial side of the lateral condyle of the femur.

  - The **POSTERIOR CRUCIATE LIGAMENT (PCL)** arises from the posterior intercondylar area of the tibia and passes anteriorly to attach to the lateral side of the medial condyle of the femur.

  - **Both cruciate ligaments function as restraining straps to prevent slipping movements at the knee joint.**

    - The ACL <u>prevents anterior sliding of the tibia</u>.

    - The PCL <u>prevents anterior sliding of the femur or backward displacement of the tibia</u>.

  - **Both the ACL and PCL function together to lock the knee when it is extended.**

PRACTICE QUESTIONS 2:
The Integumentary and Skeletal Systems

1. Under high environmental temperatures and humidity,
    a. one feels increasingly energetic as the skin surface becomes warmer.
    b. bodies are cooled primarily by the merocrine sweat glands.
    c. one can become hydrated from absorbing water vapor from the air.
    d. bodies are cooled primarily by the apocrine sweat gland.
    e. A and B

2. As keratinocytes are pushed towards the stratum corneum, they:
    a. become more viable.
    b. undergo extensive mitotic division.
    c. receive more robust nourishment.
    d. become less viable.
    e. secrete more melanin.

3. Which combination of the following is *not* a function of sebum?
    (1) lubricating the skin
    (2) cooling the skin
    (3) inhibiting the growth of bacteria on the skin
    (4) sensing pressure on the surface of the skin
    (5) promoting the production of new skin cells

    a. 2, 4, 5
    b. 1, 3
    c. 1, 2, 3
    d. 1, 2
    e. 4, 5

4. The integument is separated from the deep fascia by the:
    a. epidermis.
    b. reticular layer of the dermis.
    c. cutaneous membrane.
    d. papillary layer of the dermis.
    e. hypodermis.

5. Specialized sensory receptors in the skin include all of the following *except*:
    a. lamellated corpuscles.
    b. cells of Langerhans.
    c. Ruffini corpuscles.
    d. Merkel cells.
    e. tactile corpuscles.

6. The epidermis in a section of thick skin includes the following five layers. In what order do these occur, from the basement membrane to the surface?
    (1) Stratum granulosum ~3~
    (2) Stratum lucidum ~4~
    (3) Stratum corneum ~5~
    (4) Stratum basale (germinativum) ~1~
    (5) Stratum spinosum ~2~

    a. 1, 2, 3, 4, 5
    b. 1, 3, 5, 4, 2
    c. 4, 5, 1, 2, 3
    d. 5, 4, 3, 2, 1
    e. 3, 2, 5, 4, 1

7. Which of the following vertebral column levels has an intervertebral disc?
    a. between T₁₂ and L₁
    b. between C₇ and T₁
    c. between C₁ and C₂
    d. All of the above.
    e. A and B only

8. A small, synovial fluid-filled pocket in connective tissue, which may be connected to a joint cavity is a:
    a. bursa.
    b. syndesmosis.
    c. meniscus.
    d. gomphosis.
    e. retinaculum.

9. Pronation and supination occur at the:
    a. hip.
    b. shoulder.
    c. proximal radioulnar joint.
    d. elbow.
    e. proximal tibiofibular joint.

10. Which of the following is most unique to the knee?
    a. joint capsule
    b. ligaments
    c. bursae
    d. menisci
    e. tendons

11. Ribs 1-10 articulate with the bodies of the thoracic vertebrae and the:
    a. transverse processes of the thoracic vertebrae.
    b. spinous processes of the lumbar vertebrae.
    c. manubrium of the sternum.
    d. spinous processes of the cervical vertebrae.
    e. A and C

12. Structures that connect muscle to bone are called:
    a. menisci.
    b. ligaments.
    c. bursae.
    d. tendons.
    e. cartilage.

13. The olecranon is a part of the:
    a. radius.
    b. clavicle.
    c. scapula.
    d. humerus.
    e. ulna.

14. Which of the following vertebral structures can most easily be seen or felt on the <u>ventral</u> surface?
    a. spinous process
    b. intervertebral disc
    c. transverse process
    d. body (*centrum*)
    e. pedicle

15. The squamous suture is the boundary between which bones?
    a. parietal and temporal
    b. nasal and vomer
    c. frontal and nasal
    d. frontal and parietal
    e. frontal and sphenoid

16. The type of joint formed by the teeth in gums is:
    a. a synostosis.
    b. a symphysis.
    c. a gomphosis.
    d. a synchondrosis.
    e. None of the above.

17. Which of the following is *not* a structure of the scapula?
    a. spinous process.
    b. glenoid cavity.
    c. spinous fossa.
    d. olecranon.
    e. acromion.

18. The structure that houses the pituitary gland is found in which bone?
    a. lacrimal
    b. ethmoid
    c. parietal
    d. sphenoid
    e. palatine

19. In the ankle, the tarsal bone(s) that articulate(s) with both leg bones is/are the:
    a. cuboid bone.
    b. calcaneus.
    c. medial cuneiform bone.
    d. talus.
    e. A and D

20. The structure of the *os coxa* that can be felt on the anterior of the hip is the:
    a. ischial spine.
    b. sciatic notch.
    c. anterior superior iliac spine.
    d. anterior inferior iliac spine.
    e. pubic symphysis.

21. The patella:
    a. is normally on the lateral aspect of the knee joint.
    b. is preformed in cartilage during development.
    c. forms part of a ball-and-socket joint.
    d. increases the contraction force of the *quadriceps femoris*.
    e. C and D

22. The cruciate ligaments attach to the _____ of the tibia.
    a. epicondyles
    b. intercondylar eminence
    c. anterior tuberosity
    d. olecranon
    e. linea aspera

23. The fibula:
    a. articulates with the femur.
    b. forms an important part of the knee joint.
    c. provides lateral stability to the ankle.
    d. is the lateral bone of the leg.
    e. C and D

24. Ligaments inside a joint capsule are called:
    a. intercapsular.
    b. cruciates.
    c. extracapsular.
    d. intracapsular.
    e. None of the above.

25. The pectoral girdle meets the axial skeleton where:
    a. the clavicle attaches to the manubrium.
    b. the scapula attaches to the clavicle.
    c. the scapula attaches to the ribs.
    d. the humerus attaches to the manubrium.
    e. The appendicular skeleton and the axial skeleton do not articulate.

26. A cartilaginous joint that does *not* permit movement is a:
    a. synchondrosis.   d. synostosis.
    b. syndesmosis.   e. A and D
    c. gomphosis.

27. Most of the anterior surface of the scapula is formed by the:
    a. subscapular fossa.
    b. infraspinous fossa.
    c. supraspinous fossa.
    d. coracoid process.
    e. olecranon process.

28. All of the following are true of sutures *except*:
    a. they occur only between the bones of the skull.
    b. they may become completely fused later in life.
    c. they are synarthroses. ~ allow no movement
    d. the edges of the bones forming this joint interlock.
    e. There are no exceptions; all of the above are true.

29. Which layer is found only in the skin of the palms and the soles of the feet?
    a. stratum corneum   d. stratum lucidum
    b. stratum granulosum   e. stratum spinosum
    c. stratum basale

30. Which integumentary layer is comprised of loose areolar connective tissue?
    a. keratohyaline layer
    b. papillary layer of dermis
    c. stratum granulosum
    d. stratum basale
    e. reticular layer of dermis

31. Which are macrophage-like dendritic cells, which use receptor-mediated endocytosis to remove foreign proteins in the epidermis?
    a. Merkel cells
    b. keratohyaline granular cells
    c. Langerhans cells — got "attack what is not needed"
    d. melanocytes
    e. keratinocytes

32. In bony tissue, which structures connect the vascular and nervous supply of the periosteum to the medullary cavity?
    a. lamellae
    b. perforating (Volkmann's) canals
    c. lacunae
    d. central (Haversian) canals
    e. canaliculi

33. Which structure forms the anterior border of the *sella turcica*?
    a. clinoid process   d. optic groove
    b. dorsum sellae   e. tuberculum sellae
    c. optic canal

34. Which structure, also known as the lateral condyle of the humerus, articulates with the radius?
    a. capitulum   d. greater tubercle
    b. radial groove   e. lesser tubercle
    c. trochlea

35. Which of the following is a clavicular structure that functions as an attachment site for a ligament?
    a. greater tubercle   d. conoid tubercle
    b. coronoid tubercle   e. coracoid tubercle
    c. deltoid tuberosity

36. Which structure is located at the Y-shaped junction of the ilium, ischium, and pubis?
    a. bursa   d. acetabulum
    b. glenoid cavity   e. greater trochanter
    c. antecubital fossa

Δ shape   ilium   pubis   ischum

Chapter 9 – THE MUSCULAR SYSTEM I: SKELETAL MUSCLE TISSUE and MUSCLE ORGANIZATION

A. | **OVERVIEW OF MUSCLE TISSUE** |

- There are **3 TYPES** of muscle tissue:

    - **SKELETAL muscle**
        - Attaches to skeleton
        - Has single, very long and cylindrical striated cells
        - Multinucleate (many peripherally located nuclei)
        - Can be CONTROLLED VOLUNTARILY

    - **CARDIAC muscle**
        - Occurs in the heart wall
        - Has branching chains of striated cells
        - One nucleus per cell; some can be binucleate
        - Intercalated discs contain several types of cell junctions
        - Cardiac cells are electrically coupled by gap junctions
        - CONTROLLED INVOLUNTARILY

    - **SMOOTH muscle**
        - Occurs chiefly in walls of hollow organs
        - Single, fusiform nonstriated cells
        - Uninucleate
        - CONTROLLED INVOLUNTARILY

- The 3 types of muscle tissues share **four specialized properties**:

    1. **Excitability**: the ability of muscle cells to respond to nerve signals or other stimuli, causing electrical impulses to travel along the muscle cells' plasma membrane
    2. **Contractility**: the ability to generate a strong pulling force while muscle cells shorten (contract)
    3. **Elasticity**: the ability of a muscle, after being stretched (or contraction), to recoil passively to its original (or resting) length
    4. **Extensibility**: the ability to continue to contract over a range of resting lengths

- | **FUNCTIONS of SKELETAL MUSCLE TISSUE** |

    - Produces **movement**

    - **Maintains posture and body position**

    - **Supports soft tissue**

- **Regulates entrance and exit of materials** (skeletal muscles encircle openings/orifices of the digestive and urinary tracts)

- **Heat generation** – muscle contractions produce heat to maintain body temperature

- Joint stabilization (muscle tone stabilizes and strengthens joints)

- **GROSS ANATOMY of SKELETAL MUSCLE**

  - Skeletal muscle is surrounded by *epimysium*, and is comprised of bundles of muscle fascicles.

  - **MUSCLE FASCICLE**: a group or bundle of muscle fibers; surrounded by *perimysium*

  - **MUSCLE FIBER (or MYOFIBER or MUSCLE CELL)**: a highly elongated cell (in skeletal or smooth muscle, not cardiac muscle) comprised of *myofibrils*; surrounded by *endomysium*

- **MICROSCOPIC ANATOMY of SKELETAL MUSCLE FIBERS**

  - **SARCOLEMMA**: plasma membrane of muscle fiber (or myofiber or muscle cell), which is comprised of abundant *myofibrils*

  - **SARCOPLASM**: cytoplasm of muscle fiber (or muscle cell), which contains numerous *myofibrils*

  - **MYOFIBRIL**: a cylindrical structure, which is as long as the entire cell and consists of *sarcomeres*; surrounded by *sarcoplasmic reticulum*

    - Myofibrils can shorten, and are, hence, responsible for contraction of the skeletal muscle fiber.

    - Myofibril contraction leads to shortening the entire cell because a myofibril is attached to the sarcolemma at each end of the cell.

  - **SARCOPLASMIC RETICULUM (SR)**: an internal membrane complex that is similar to the smooth endoplasmic reticulum (SER) of other cells

    - Closely associated with the transverse (T) tubules

    - Plays important role in controlling the contraction of individual myofibrils via the release of calcium ions (Ca^{2+})

- **TRANSVERSE (T) TUBULES**: deep invaginations of the sarcolemma (into the sarcoplasm), which allow *electrical impulses* (that stimulate the membrane to contract) to quickly travel to the interior of the cell

- **TERMINAL CISTERNAE**: expanded chambers on either side of a transverse tubule where the tubule of the SR has enlarged and fused

- **TRIAD**: the combination of a pair of terminal cisternae plus a transverse tubule

- <u>**SARCOMERE**</u>: myofibrils consist of sarcomeres, which are repeating units of *myofilaments*; smallest functional units of muscle fiber

- <u>**MYOFILAMENTS**</u>: the sarcoplasm of muscle contains protein filaments, which generate contractile force

    o **The myofilaments determine the <u>striation pattern</u> in skeletal muscle fibers.**

    o The myofilaments fill most of the sarcoplasm (cytoplasm) of each muscle fiber.

    o They are organized in the repeating units called <u>sarcomeres</u>.

    o There are two primary types of myofilaments in muscle cells:

        ▪ **ACTIN**: protein filaments found in THIN filaments

        ▪ **MYOSIN**: protein filaments found in THICK filaments

- **SARCOMERE STRUCTURE / ORGANIZATION**

    The striated (banded) appearance of skeletal muscle tissue arises from the organization of the thick and thin filaments within the myofibrils of the sarcomere.

    o **MYOFIBRILS** are arranged **parallel to the long axis** of the cell, with their **sarcomeres arranged side to side**, giving the appearance of **distinct dark and light bands**. Hence, the striations correspond to these bands of the individual sarcomeres.

    o **M LINE:** a group of proteins, which link the thick filaments that lie in the center of the sarcomere

    o **Z LINES (or Z DISCS):** open meshwork of <u>interconnecting proteins</u> called **ACTININS**, which occur where thin filaments from adjacent sarcomeres join

o   The thin filaments are attached to the Z line, and extend toward the M line.

o   **ZONE of OVERLAP:** area where the <u>thin filaments pass between the thick filaments</u>

    ■   In 3-D cross-sectional view: each thin filament is surrounded by three thick filaments, arranged in a triangle, while six thin filaments surround each thick filament.

o   **A BAND** (*anisotropic band*): the area containing thick filaments, including the M line, the H band, and the zone of overlap; appears as a *"dark band"*

o   **H BAND:** area containing THICK filaments only

o   **I BAND** (*isotropic band*): the area containing THIN filaments only; appears as a *"light band"*

o   <u>During contraction</u>, the Z lines/discs move closer together, and the I bands and H bands shorten.

---

*Memory aid: Remember that <u>**A**</u> **bands are** *<u>d</u>ark* and <u>**I**</u> **bands are** *light* when viewed under polarized light; hence, the *<u>a</u>nisotropic* and *<u>i</u>sotropic* characteristics assigned to them, respectively.

---

o   **THIN FILAMENTS:**

    ■   **F Actin:** a strand of 300-400 globular G actin molecules

    ■   **NEBULIN:** a slender strand of proteins that holds the F actin strand together

    ■   **G Actin** molecule: contains an **active site** that can bind to a thick filament, in the same manner that a substrate molecule binds to an enzyme's active site

    ■   **TROPOMYOSIN:** protein molecules that form a long chain, which covers the active sites, preventing actin-myosin interaction

    ■   **TROPONIN:** protein molecules that hold the tropomyosin strand in place

        •   Changes position to move the tropomyosin molecule, exposing the active site, prior to a muscle contraction

        •   **Acts as the regulator molecule of a muscle contraction**

- o **THICK FILAMENTS:**

  - ▪ **MYOSIN:** approximately 500 myosin molecules comprise the thick filament

    - • Each myosin molecule consists of a double myosin strand with an attached, elongate *tail* and a free globular *head*.

  - ▪ **Proteins of the M line** <u>interconnect adjacent thick filaments</u>, midway along their length.

  - ▪ **Myosin heads are also known as CROSS-BRIDGES** because they connect thick filaments and thin filaments during a muscle contraction.

  - ▪ Myosin heads project outward toward the surrounding thin filaments, as the entire myosin molecules are oriented away from the M line.

  - ▪ **TITIN:** protein molecule that <u>makes up the core</u> of each **thick** filament

    - ▪ A strand of titin (on either side of the M line) <u>extends the length of the filament</u> and **attaches the M line to the Z line** (or Z disc).

    - ▪ An exposed portion within the I band is **highly elastic** and will recoil after stretching.

    - ▪ Completely relaxed in the normal resting sarcomere

    - ▪ Becomes tense only when some external force stretches the sarcomere

    - ▪ <u>When sarcomere is stretched</u>, the titin strands **help maintain the normal alignment of the thick and thin filaments.**

    - ▪ <u>When the stretching force is removed</u>, titin fibers **help return the sarcomere to its normal resting length.**

B. FOCUS ON SKELETAL MUSCLE

- • **BASIC FEATURES of a SKELETAL MUSCLE**

  - o VARIOUS SHAPES: **spindle-shaped cylinders / triangles / sheets**

  - o <u>Each skeletal muscle is an organ.</u> It contains connective tissue elements, blood vessels and nerves, as well as being comprised mostly of muscle fibers.

  - o <u>Organization levels</u> in a skeletal muscle (gross anatomy to microscopic anatomy level):

- WHOLE MUSCLE - FASCICLE – FIBER (or *myofiber* or muscle cell) - **MYOFIBRIL - SARCOMERE - MYOFILAMENT (Actin and Myosin)**

  o <u>Connective tissue elements</u>:

    ▪ **EPIMYSIUM**: dense irregular connective tissue (CT) sheath wrapped <u>around a whole muscle</u>

    ▪ **PERIMYSIUM**: fibrous CT sheath <u>around a fascicle</u>

    ▪ **ENDOMYSIUM**: thin reticular fiber CT sheath <u>around each muscle fiber</u>

  o Every skeletal muscle fiber is innervated and stimulated by a nerve cell to contract; has a rich blood supply; and has fine nerve fibers and capillaries in the endomysium.

  o **ORIGIN**: immovable (or less movable) attachment from which a muscle extends

  o **INSERTION**: more movable attachment

  o <u>Skeletal muscles attach to bones through</u> **TENDONS, APONEUROSES,** or **DIRECT (fleshy) ATTACHMENTS**. (<u>Bones</u> connect to other bones via LIGAMENTS.)

- | **FUNCTIONAL ANATOMY of SKELETAL MUSCLE TISSUE** |

  - A **MUSCLE CONTRACTION** exerts a pull, or **tension**, and <u>shortens</u> the muscle fiber in length.

    - The muscles attach to the skeleton in a way that keeps them at a **near-optimal length for generating maximum contractile forces**.

      o This is accurate for muscles involved in activities such as walking, in which the muscles contract and relax repeatedly. <u>The muscle fibers are stretched to near-optimal length before stimulation of contraction occurs.</u>

    - The presence of **calcium ions (Ca^{2+})** is the **trigger** for a contraction.

    - The presence of **ATP** is **required** for the contraction to occur.

- | **SLIDING FILAMENT THEORY** or **MECHANISM** | : theory or mechanism that explains the physical changes that occur between the thin and thick filaments during muscle contraction
    - The H band and I band get smaller

- The zone of overlap gets larger
- The Z lines move closer together
- The width of the A band remains constant throughout the contraction

- **MECHANISM STEPS of the SLIDING FILAMENT THEORY:**
  - **Myosin heads** of thick filaments **bind** to active sites on thin filaments, causing sliding to occur.
  - **CROSS-BRIDGE BINDING:** the myosin head **pivots** toward the M line, pulling the thin filament toward the center of the sarcomere
  - **Cross-bridge** then **detaches** and **returns** to its original position. It is then ready for the next cycle of "bind, pivot, detach, and return" regarding its original position.
  - The **Z lines move toward the M line** when the thick filaments pull on the thin filaments, **causing the sarcomere to shorten.**

- ☐ **NEURAL CONTROL of MUSCLE FIBER CONTRACTION** ☐

- When a nerve cell stimulates a muscle fiber, it sets up an impulse in the sarcolemma that signals the sarcoplasmic reticulum to release Ca^{2+}, which then initiates the sliding of the myofilaments. This event translates to a muscle contraction.

- ☐ **MOTOR NEURONS innervate individual skeletal muscle fibers** ☐ at **neuromuscular synapses or junctions (motor end plates).**

  - **NEUROMUSCULAR SYNAPSE:** a specific synapse between a motor neuron and a muscle cell

  - **SYNAPTIC TERMINAL:** the expanded tip of the motor neuron's axon, at the neuromuscular synapse

  - **SYNAPTIC VESICLES:** small secretory vesicles (filled with acetylcholine) in the cytoplasm of the synaptic terminal

  - **NEUROTRANSMITTER:** a chemical released by a neuron to communicate with another cell

  - **ACETYLCHOLINE (ACh):** a neurotransmitter that signals the muscle cell to contract; released at the axon terminal.

  - **SYNAPTIC CLEFT:** a narrow space that separates the synaptic terminal from the motor end plate of the skeletal muscle fiber

  - **ACETYLCHOLINESTERASE (AChE)** or *cholinesterase*: an enzyme, which breaks down ACh molecules, and is released (by the basal lamina of the cell) into the synaptic cleft

- **EVENTS during NEURAL STIMULATION of a MUSCLE:**

  o An <u>electrical impulse</u> arrives at the synaptic terminal.

  o **ACh is released** and **binds to receptor sites** on motor end plate.

  o A change in local transmembrane potential occurs, resulting in generation of an electrical impulse, or **action potential (AP)**.

  o Electrical impulse (or AP) travels <u>all over the surface of the sarcolemma and into each T tubule.</u>

  o **APs continue to be generated until AChE removes the bound ACh.**

    o Immediately after ACh signals a single contraction, it is broken down by AChE.

- <u>Each muscle fiber must be served by a neuromuscular junction.</u>

- **MOTOR UNIT** : consists of one motor neuron and all the skeletal muscle fibers it innervates (controls)

  - Contains different numbers of muscle fibers distributed widely within a muscle

  - All muscle fibers in a motor unit contract simultaneously.

  - The size of a motor unit indicates the level of control of the movement.

    o A motor neuron that controls two or three muscle fibers, such as those in the muscles of the eye, has more precise control of muscle movement than a motor neuron that innervates up to 2000 muscle fibers, such as those in leg muscles, which require much less precise control.

- **MOTOR UNITS and MUSCLE CONTROL**

  - **MUSCLE TWITCH:** a single, momentary contraction, which is a response to a single stimulus

    - Contain different numbers of muscle fibers distributed widely within a muscle

  - **ALL-or-NONE PRINCIPLE:** a characteristic in which each muscle fiber either contracts completely or does not contract at all

- All muscle fibers in a motor unit contract simultaneously.

- The amount of force, exerted by the muscle as a whole, depends on how many motor units are activated.

- **RECRUITMENT (or *multiple motor unit summation*):** the smooth, but steady, increase in muscular tension that is produced by increasing the number of motor units that is activated

- **MUSCLE TONE:** the resting tension in a skeletal muscle

  - In a resting muscle, some motor units are always active without producing enough tension to cause movement, but the activity tenses the muscle.

  - Stabilizes the position of bones and joints

  - **MUSCLE SPINDLES:** specialized muscle cells that are monitored by sensory nerves, which control the muscle tone in the surrounding muscle tissue

- **MUSCLE HYPERTROPHY:** enlargement of skeletal muscles that <u>undergo excessive repeated stimulation</u> that produces near-maximal tension

- **MUSCLE ATROPHY:** reduction in skeletal muscle size, tone, and power as <u>a result of inadequate stimulation to maintain resting muscle tone</u> in the affected area

- **TYPES of SKELETAL MUSCLE FIBERS**

  - **Slow oxidative fibers, or *red fibers* (Type I fibers)**

    o Relatively thin fibers, which are red because of their **abundant *myoglobin*** (oxygen-binding pigment in sarcoplasm)

    o Obtain energy from aerobic metabolic reactions (hence, abundant mitochondria and rich capillary supply)

    o Contract slowly; extremely resistant to fatigue as long as enough oxygen is present

    o Deliver prolonged contractions

    o **Best for maintaining postures**

- ☐ **Intermediate (fast oxidative) fibers (Type IIa fibers)**

  - ○ Contract quickly

  - ○ Oxygen dependent and have a high myoglobin content, abundant mitochondria and rich capillary supply

  - ○ Fatigue resistant but less so than Type I

  - ○ Intermediate in many of their characteristics compared to Types I and IIx

  - ○ **Best for long-term production of fairly strong contractions, such as required in locomotion of the lower limbs**

- ☐ **Fast glycolytic fibers, or *white fibers* (Type IIx fibers)**

  - ○ Pale fibers because they contain **little *myoglobin***

  - ○ About twice the diameter of Type I fibers

  - ○ Contain more myofilaments

  - ○ Generate much more power

  - ○ Depend on **anaerobic pathways** to make ATP

    - ○ Hence, few mitochondria or capillaries are present, but there are **abundant glycosomes** that contain glycogen as a fuel source.

  - ○ Contract rapidly; fatigue quickly

  - ○ **Best for short bursts of power**, such as required in lifting heavy objects for brief periods (muscles of the upper limbs)

- ☐ **LEVER SYSTEMS: BONE-MUSCLE RELATIONSHIPS**

  - **LEVER**: a rigid bar or structure that moves on a fixed point, called the **FULCRUM**

  - When effort is applied to the lever – a load is moved.

  - **In the human body:**
    - ○ **BONES = LEVERS**
    - ○ **JOINT = FULCRUM**
    - ○ **EFFORT is exerted by skeletal muscles pulling on their insertions.**

- **CLASSES OF LEVERS**

  - **1st class levers** (effort-fulcrum-load): may operate at a <u>mechanical</u> <u>advantage</u> or <u>disadvantage</u>

  - **2nd class levers** (fulcrum-load-effort): all work at a <u>mechanical advantage</u>

  - **3rd class levers** (fulcrum-effort-load): all work at a <u>mechanical disadvantage</u>

  *Most muscles of the body are in 3rd class lever systems to provide speed of movement.

- **MECHANICAL ADVANTAGE** (effort is farther from the fulcrum than is the load): **allows slow and strong movements**

- **MECHANICAL DISADVANTAGE** (effort is closer to the fulcrum): **allows muscles to move quickly and far but takes extra effort**

- **ORGANIZATION OF FASCICLES IN SKELETAL MUSCLES**
  (Fascicle = bundle of muscle fibers)

- **PARALLEL muscle** – the long axes of the fascicles are *parallel* to the long axis of the muscle, and the muscle extends from origin to insertion

  - Most of the skeletal muscles in the body are parallel muscles.

  - Such a muscle has a central ***body***, or ***belly***, or ***gaster***.

  - *Biceps brachii muscle* of the arm is an example.

- **CONVERGENT muscle** – the origin is broad, and the fascicles *converge* toward the tendon of insertion, its common attachment site

  - The fibers may pull on a <u>tendon</u>, a <u>tendinous sheet</u>, or a slender band of collagen fibers known as a <u>raphe</u>.

  - *Pectoralis major muscle* of the anterior chest is an example.

- **CIRCULAR muscle** (or ***sphincter***) – the fascicles are arranged in concentric rings around an opening or recess

  - When the muscle contracts, the diameter of the opening decreases.

  - *Orbicularis oris muscle* of the mouth is an example.

- **PENNATE muscle** (*penna*, feather) – the fascicles are short and attach at an oblique angle to a tendon that runs through the body (the whole length) of the muscle

  o **UNIPENNATE** – the fascicles insert into one side of the tendon

    ▪ *Extensor digitorum longus muscle* of the anterior leg

  o **BIPENNATE** – the fascicles insert into both sides of the tendon

    ▪ *Rectus femoris muscle* of the thigh

  o **MULTIPENNATE** – the fascicles insert into a tendon that branches within the muscle

    ▪ Arrangement looks like many feathers situated side by side, whose quills are all inserted into one tendon

    ▪ *Deltoid muscle* of the shoulder

- ┌───┐
  │ **INTERACTIONS OF SKELETAL MUSCLES IN THE BODY** │
  └───┘

  - Skeletal muscles are arranged in opposing groups across movable joints, allowing one group of muscles to reverse or modify the action of the opposing group.

  - **AGONIST (or prime mover):** a muscle whose contraction bears the main responsibility FOR A PARTICULAR MOVEMENT

    o e.g., ***biceps brachii*** for <u>forearm flexion</u> at the elbow

  - **ANTAGONIST:** group(s) of muscles whose actions oppose that of the corresponding agonist

    o e.g. ***triceps brachii***, during forearm flexion, are stretched and maybe slightly relaxed, and stabilizes the flexion movement; BUT THEY ALSO ACT AS AGONISTS for <u>extension</u> of the forearm

┌──┐
│ ***REMEMBER THAT A MUSCLE CANNOT BE CATEGORIZED AS AN AGONIST**   │
│ **ONLY OR AS AN ANTAGONIST ONLY.** WHETHER OR NOT A MUSCLE IS   │
│ TO BE CONSIDERED AGONIST vs. ANTAGONIST, IT ALL **DEPENDS ON THE**   │
│ **PARTICULAR MOVEMENT TO BE EXECUTED.**   │
└──┘

  - **SYNERGISTS:** aid the agonists, either by adding a little extra force to the same movement or by reducing undesirable extra movements that the agonist may produce; they stabilize joints, as ***fixators***.

- ☐ **NAMING THE SKELETAL MUSCLES according to the following CRITERIA**

  - **LOCATION** of the muscle

    - *Brachialis muscle* in the arm (*brachium* = arm)

    - *Intercostal muscles* lie between the ribs (*costa* = rib)

    - ***Externus*** (external), or ***superficialis*** (superficial) – describe muscles that are visible at the body surface

    - ***Internus*** (internal), or ***profundus*** – describe muscles lying beneath the body surface

    - ***Extrinsic*** – describes superficial muscles that position or stabilize an organ

    - ***Intrinsic*** - describes deep muscles that function within the organ

  - **SHAPE** of the muscle

    - ***Trapezius*** – trapezoidal shape

    - ***Deltoid*** – triangular

    - ***Rhomboideus*** – a rhomboid

    - ***Orbicularis*** – a circle

  - **RELATIVE SIZE of the muscle:**

    - ***Magnus*** (large), ***major*** (larger) or ***maximus*** (largest) vs. ***minor*** (small) or ***minimus*** (smallest);

    - ***Longus*** (long) or ***longissimus*** (longest) vs. *brevis* (short)

    - ***Teres*** (long and round)

  - **DIRECTION OF FASCICLES and the MUSCLE FIBERS:**

    - ***Rectus*** (straight)

    - ***Transversus*** (fascicles lie at right angles)

    - ***Oblique*** (fascicles lie at oblique angles)

- **LOCATION OF ATTACHMENTS:** The names reveal points of origin and insertion.

  o The ***origin*** is <u>always named first</u>.

- **NUMBER OF ORIGINS:**

  o ***Biceps*** (2 origins)

  o ***Triceps*** (3 origins)

  o ***Quadriceps*** (4 origins)

- **ACTION** of the muscle:

  o ***Flexor*** vs. ***extensor***

  o ***A<u>b</u>ductor*** vs. ***adductor***

---

*Except for the ***platysma*** and the ***diaphragm***, the complete names of all skeletal muscles include the term *muscle*, but textbooks often choose to drop this term for the sake of brevity.

Chapter 10 – THE MUSCULAR SYSTEM II: MAJOR MUSCLES of the AXIAL SKELETON

Study Tips: It is best to be familiar with the muscle names, description and actions of each in the following sections. For introductory-intermediate courses, <u>exam questions</u> typically involve identification of the muscles, based on description, actions and origin/insertion.

<u>In lab exams</u>, identification of muscles, by sight, is a common approach. <u>In lecture exams</u>, especially multiple-choice question types, questions often involve identifying muscles if given only the <u>origin and insertion</u>, or if given the <u>action, description and/or location of the muscle</u>; or the reverse is possible in which an action, origin, or insertion is asked if given the name of the muscle. Question types and content will always vary according to the professor's preference.

Remember these four aspects of studying muscles: the muscle name, its origin, its insertion, and its major action(s). Hence, this study guide focuses on readily providing each aspect for the major muscles listed in the following sections. *<u>Primary action(s) for each muscle is/are in upper case lettering</u>.*

MUSCLES OF THE HEAD, NECK, and THORAX

(**O** = ORIGIN / **I** = INSERTION / **A** = ACTION)

A. | **MUSCLES of the HEAD: FACIAL EXPRESSION**

MUSCLES OF THE SCALP:

- **EPICRANIUS (or OCCIPITOFRONTALIS):**

  o Bipartite muscle consisting of two muscles connected by a cranial aponeurosis, called the **GALEA APONEUROTICA**; the alternate actions of the two muscles pull the scalp forward and backward

  - **FRONTALIS** muscle **(or FRONTAL BELLY):** covers the forehead and dome of skull; no bony attachments
    O – galea aponeurotica
    I – skin of eyebrows and root of nose
    A - with aponeurosis fixed, it RAISES the EYEBROWS (as in surprise); wrinkles forehead skin horizontally

  - **OCCIPITALIS** muscle **(or OCCIPITAL BELLY):** overlies the posterior occiput

O – occipital and temporal (mastoid) bones
I – galea aponeurotica
A – FIXES APONEUROSIS and PULLS THE SCALP POSTERIORLY

MUSCLES OF THE FACE:

- **CORRUGATOR SUPERCILII**: small muscle; activity associated with that of orbicularis oculi
  O – arch of frontal bone above nasal bone
  I – skin of eyebrow
  A - DRAWS EYEBROWS TOGETHER and INFERIORLY wrinkles skin of forehead vertically (as in frowning)

- **ORBICULARIS OCULI**: thin, flat sphincter muscle of eyelid; surrounds the rim of the orbit
  O – frontal and maxillary bones and ligaments around orbit
  I – tissue of eyelid
  A - CLOSES the EYE; protects from intense light and injury; produces blinking, squinting, and draws eyebrows inferiorly

- **ZYGOMATICUS**: muscle pair extending diagonally from cheekbone to corner of mouth
  O – zygomatic bone
  I – skin and muscle at corner of mouth
  A – RAISES LATERAL CORNERS of MOUTH UPWARD (SMILING MUSCLE)

- **RISORIUS**: slender muscle inferior and lateral to zygomaticus
  O – lateral fascia associated with masseter muscle
  I – skin at angle of mouth
  A - DRAWS CORNER OF LIP LATERALLY; tenses lips; <u>synergist</u> of zygomaticus

- **LEVATOR LABII SUPERIORIS**: thin muscle between orbicularis oris and inferior eye margin
  O – zygomatic bone and infraorbital margin of maxilla
  I – skin and muscle of upper lip
  A - OPENS LIPS; raises and furrows the upper lip

- **DEPRESSOR LABII INFERIORIS**: small muscle running from mandible to lower lip
  O – body of mandible lateral to its midline
  I – skin and muscle of lower lip
  A - DRAWS LOWER LIP INFERIORLY (as in a pout)

- **ORBICULARIS ORIS**: complicated, multilayered muscle of the lips with fibers that run in different directions, mostly circularly

O – arises indirectly from maxilla and mandible; fibers blended with fibers of other facial muscles associated with the lips
I – encircles mouth; inserts into muscle and skin at angles of mouth
A - CLOSES, purses and protrudes LIPS; kissing and whistling muscle

- **MENTALIS:** one of the muscle pair forming a V-shaped muscle mass on chin
  O – mandible below incisors
  I – skin of chin
  A – protrudes lower lip; WRINKLES CHIN

- **BUCCINATOR:** thin, horizontal (principal) cheek muscle; deep to masseter
  O – molar region of maxilla and mandible
  I – orbicularis oris
  A - draws corner of mouth laterally; COMPRESSES CHEEK (as in whistling and sucking); holds food between teeth during chewing; well developed in nursing infants

- **PLATYSMA:** unpaired, thin, sheet-like superficial neck muscle
  O – fascia of chest (over pectoral muscles and deltoid)
  I – lower margin of mandible; and skin and muscle at corner of mouth
  A – helps depress mandible; pulls lower lip back and down, i.e., produces downward sag of mouth; TENSES SKIN of NECK (e.g., during shaving)

B. | **MUSCLES of the HEAD: MASTICATION** |

- **MASSETER:** powerful muscle that covers the lateral aspect of the mandibular ramus
  O – zygomatic arch and zygomatic bone
  I – angle and ramus of mandible
  A - PRIME MOVER of JAW CLOSURE; elevates mandible

- **TEMPORALIS:** fan-shaped muscle that covers parts of the temporal, frontal and parietal bones
  O – temporal fossa
  I – coronoid process of mandible via a tendon that passes deep to zygomatic arch
  A - CLOSES JAW; elevates and retracts mandible; maintains position of the mandible at rest; deep anterior part may help protract mandible

- **MEDIAL PTERYGOID:** deep two-headed muscle that runs along internal surface of mandible, and it largely concealed by that bone
  O – medial surface of lateral pterygoid plate of sphenoid bone, maxilla, and palatine bone
  I – medial surface of mandible near its angle

A - acts <u>with lateral pterygoid muscle</u> to PROTRACT MANDIBLE and to PROMOTE SIDE-TO-SIDE (GRINDING) MOVEMENTS; synergist of temporalis and masseter muscles in elevation of the mandible

- **LATERAL PTERYGOID:** deep two-headed muscle; lies superior to medial pterygoid muscle
  O – greater wing and lateral pterygoid plate of sphenoid bone
  I – condyle of mandible and capsule of temporomandibular joint
  A - PROVIDES FORWARD SLIDING and SIDE-TO-SIDE GRINDING MOVEMENTS of the LOWER TEETH; protracts mandible (pulls it anteriorly)

- **BUCCINATOR:** thin, horizontal (principal) cheek muscle
  O – molar region of maxilla and mandible
  I – orbicularis oris
  A – COMPRESSES the CHEEK; helps keep food between grinding surfaces of teeth during chewing

C. | MUSCLES PROMOTING TONGUE MOVEMENTS | (EXTRINSIC MUSCLES)

- **GENIOGLOSSUS:** fan-shaped muscle; forms bulk of inferior part of tongue; its attachment to mandible prevents tongue from falling backward and obstructing respiration
  O – internal surface of mandible near symphysis
  I – inferior aspect of the tongue and body of hyoid bone
  A – PRIMARILY PROTRACTS TONGUE, but can depress or act in concert with other extrinsic muscles to retract tongue

- **HYOGLOSSUS:** flat, quadrilateral muscle
  O – body and greater horn of hyoid bone
  I – inferolateral tongue
  A – DEPRESSES TONGUE and draws its sides downward

- **STYLOGLOSSUS:** slender muscle running superiorly to and at right angles to hyoglossus
  O – styloid process of temporal bone
  I – lateral inferior aspect of tongue
  A – RETRACTS and ELEVATES TONGUE

D. | MUSCLES of the ANTERIOR NECK and THROAT | : SWALLOWING and SPEAKING

The neck is divided into ANTERIOR and POSTERIOR TRIANGLES by the STERNOCLEIDOMASTOID muscle. The ANTERIOR TRIANGLE is divided into the SUPRAHYOID MUSCLE GROUP and the INFRAHYOID MUSCLE GROUP.

SUPRAHYOID MUSCLES : muscles that help form floor of oral cavity, anchor tongue, elevate hyoid, and move larynx superiorly during swallowing; lie superior to hyoid bone

- **DIGASTRIC:** consists of two bellies united by an intermediate tendon, forming a V shape under the chin
  O – lower margin of mandible (anterior belly) and mastoid process of the temporal bone (posterior belly)
  I – by a connective tissue loop to hyoid bone
  A - OPEN MOUTH and DEPRESS MANDIBLE; acting in concert, the digastric muscles elevate the hyoid bone and steady it during swallowing and speech

- **STYLOHOID:** slender muscle below angle of jaw; parallels posterior belly of digastric muscle
  O – styloid process of temporal bone
  I – hyoid bone
  A - ELEVATES and RETRACTS HYOID, thereby elongating floor of mouth during swallowing

- **MYLOHYOID:** flat, triangular muscle just deep to digastric muscle; this muscle pair forms a sling that forms the floor of the anterior mouth
  O – medial surface of mandible
  I – hyoid bone and median raphe
  A - ELEVATES HYOID BONE and FLOOR of MOUTH, enabling tongue to exert backward and upward pressure that forces food bolus into pharynx

- **GENIOHYOID:** narrow muscle in contact with its partner medially; runs from chin to hyoid bone deep to mylohyoid
  O – inner surface of mandibular symphysis
  I – hyoid bone
  A - PULLS HYOID BONE SUPERIORLY and ANTERIORLY, shortening floor of mouth and widening pharynx for receiving food during swallowing

INFRAHYOID MUSCLES : straplike muscles that depress the hyoid bone and larynx during swallowing and speaking

- **STERNOHYOID** (belongs to the infrahyoid muscle group): most MEDIAL MUSCLE of the neck; thin; superficial except inferiorly, where it is covered by the sternocleidomastoid muscle
  O – manubrium and medial end of clavicle
  I – lower margin of hyoid bone
  A - DEPRESSES LARYNX and HYOID BONE <u>if mandible is fixed</u>; may also flex skull

- **STERNOTHYROID:** lateral and deep to sternohyoid
  O – posterior surface of manubrium of sternum
  I – thyroid cartilage
  A - PULLS LARYNX and HYOID BONE INFERIORLY

- **OMOHYOID:** straplike muscle with two bellies united by an intermediate tendon; lateral to sternohyoid
  O – superior surface of scapula
  I – hyoid bone, lower border
  A - DEPRESSES and RETRACTS HYOID BONE

- **THYROHYOID:** appears as a superior continuation of sternothyroid muscle
  O – thyroid cartilage
  I – hyoid bone
  A - DEPRESSES HYOID BONE or ELEVATES LARYNX <u>if hyoid bone is fixed</u>

E. | MUSCLES of the NECK and VERTEBRAL COLUMN | : HEAD MOVEMENTS and TRUNK EXTENSION

| ANTEROLATERAL NECK MUSCLES |

- **STERNOCLEIDOMASTOID** (Anterolateral neck muscle): two-headed muscle located deep to platysma; key muscular landmark in neck
  O – manubrium of sternum and medial portion of clavicle
  I – mastoid process of temporal bone and superior nuchal line of occipital bone
  A – FLEXES and LATERALLY ROTATES the HEAD; simultaneous contraction of both muscles causes neck flexion, generally against resistance as when one raises head when lying on back; acting alone, each muscle rotates head toward shoulder on opposite side and tilts or laterally flexes head to its own side

- **SCALENES:** located more laterally than anteriorly on neck; deep to platysma and sternocleidomastoid
  O – transverse processes of cervical vertebrae
  I – anterolaterally on first two ribs
  A - ELEVATE FIRST TWO RIBS (aid in inspiration); flex and rotate neck

| INTRINSIC MUSCLES OF THE VERTEBRAL COLUMN |

SUPERFICIAL LAYER:

- **SPLENIUS:** broad bipartite superficial muscle (*capitis* and *cervicis* parts) extending from upper thoracic vertebrae to skull;

O – ligamentum nuchae, spinous processes of C_7-T_6

I – mastoid process of temporal bone and occipital bone (*capitis*); transverse processes of C_2 - C_4 vertebrae (*cervicis*)

A - EXTEND or HYPEREXTEND HEAD; when splenius muscles on one side are activated, head is rotated and bent laterally toward same side

-      ERECTOR SPINAE   (Intrinsic muscles of the back): PRIME MOVER of BACK EXTENSION; these muscles, on each side, consist of three columns of muscles, **whose subset muscles are named according to their location on the vertebral column**

  - **ILIOCOSTALIS** *(cervicis, thoracis, lumborum regions)* **GROUP**: MOST LATERAL MUSCLE GROUP of erector spinae, extending from pelvis to neck
    O – iliac crests (*lumborum*); inferior 6 ribs (*thoracis*); ribs 3 to 6 (*cervicis*)
    I – angles of ribs (*lumborum* and *thoracis*); transverse processes of C_6-C_4 (*cervicis*)
    A – EXTEND and LATERALLY FLEX the VERTEBRAL COLUMN; maintain erect posture; acting on one side, bend vertebral column to same side

  - **LONGISSIMUS** *(capitis, cervicis, thoracis regions)* **GROUP**: INTERMEDIATE TRIPARTITE MUSCLE GROUP, which extends from lumbar region to skull;
    O – transverse processes of lumbar through cervical vertebrae
    I – transverse processes of thoracic or cervical vertebrae and to ribs superior to origin as indicated by name; *capitis* inserts into mastoid process of temporal bone
    A – (*thoracis* and *cervicis*) EXTEND and LATERALLY FLEX VERTEBRAL COLUMN; acting on one side, causes lateral flexion; (*capitis*) EXTENDS HEAD and TURNS the FACE TOWARD SAME SIDE

  - **SPINALIS** *(cervicis, thoracis regions)* **GROUP**: MOST MEDIAL MUSCLE GROUP; *cervicis* usually rudimentary and poorly defined
    O – spines of upper lumbar and lower thoracic vertebrae
    I – spines of upper thoracic and cervical vertebrae
    A – EXTENDS VERTEBRAL COLUMN

    > ***Mnemonic device:*** **"*I Like Spaghetti*"** -- to remember the three columns of muscles of the erector spinae (**lateral to medial**), "**I**liocostalis, **L**ongissimus, and **S**pinalis."

DEEP MUSCLES of the SPINE (TRANSVERSOSPINALIS):

- **SEMISPINALIS** *(capitis, cervicis, and thoracis regions)*: composite muscle forming part of deep layer of intrinsic back muscles
  O – transverse processes of C_7 – T_{12}
  I – occipital bone (*capitis*) and spinous processes of cervical (*cervicis*) and thoracic vertebrae T_1 to T_4 (*thoracis*)

A – (*capitis*) <u>Together</u>, the two sides EXTEND the NECK; <u>alone</u>, each side EXTENDS and LATERALLY FLEXES NECK and TURNS HEAD to OPPOSITE SIDE; (*cervicis* and *thoracis*) EXTENDS VERTEBRAL COLUMN and ROTATES toward OPPOSITE SIDE

- **MULTIFIDUS:** relatively short muscle
  O – sacrum and transverse process of each vertebra
  I – spinous processes of the 3ʳᵈ or 4ᵗʰ more superior vertebra
  A – EXTENDS VERTEBRAL COLUMN and ROTATES toward OPPOSITE SIDE

- **INTERSPINALES:** relatively short muscle
  O – spinous process of each vertebra
  I – spinous processes of more superior vertebra
  A – EXTENDS VERTEBRAL COLUMN

- **INTERTRANSVERSARII:**
  O – transverse processes of each vertebra
  I – transverse process of more superior vertebra
  A – LATERAL FLEXION of VERTEBRAL COLUMN

- **QUADRATUS LUMBORUM:** fleshy muscle forming part of posterior abdominal wall
  O – iliac crest and lumbar fascia
  I – transverse processes of upper lumbar vertebrae / lower margin of 12ᵗʰ rib
  A - together, they depress the ribs; FLEXES VERTEBRAL COLUMN LATERALLY when acting separately; fixes floating ribs 11-12 during forced exhalation; assists in forced inhalation

F. | **DEEP MUSCLES of the THORAX: BREATHING** |

- The thoracic muscles are very short.

- Most run only from one rib to the next.

- They form three layers in the wall of the thorax.

- **EXTERNAL INTERCOSTALS:** 11 pairs lie between ribs; fibers run obliquely from each rib to rib below
  O – inferior border of rib above
  I – superior border of rib below
  A - with first ribs fixed by scalene muscles, PULL RIBS TOWARD ONE ANOTHER to ELEVATE RIB CAGE; aid in inspiration; synergists of diaphragm

- **INTERNAL INTERCOSTALS**: 11 pairs lie between ribs; fibers run deep to and at right angles to those of external intercostals (i.e., run downward and posteriorly)
  O – superior border of rib below
  I – inferior border (costal groove) of rib above
  A - with 12th ribs fixed by quadratus lumborum and abdominal muscles, the internal intercostals DRAW RIBS TOGETHER and DEPRESS RIB CAGE; aid in forced expiration; antagonistic to external intercostals

- **DIAPHRAGM:** broad muscle pierced by the aorta, inferior vena cava, and esophagus; forms floor of thoracic cavity
  O – inferior internal surface of rib cage and sternum, costal cartilages of last six ribs, and lumbar vertebrae
  I – central tendon
  A - PRIME MOVER of INSPIRATION; FLATTENS ON CONTRACTION, increasing vertical dimensions of thorax; when strongly contracted, dramatically increases intra-abdominal pressure

G. | **MUSCLES of the ABDOMINAL WALL** | : **TRUNK MOVEMENTS and COMPRESSION OF ABDOMINAL VISCERA**

Four paired flat muscles, which are very important in SUPPORTING and PROTECTING ABDOMINAL VISCERA
Other Functions: FLEXION, LATERAL FLEXION and ROTATION of the trunk; COMPRESSES ABDOMINAL CONTENTS

- **RECTUS ABDOMINIS** (*rectus* = "straight"): medial muscle pair of <u>anterior</u> wall, whose fibers extend vertically from pubis to rib cage
  - Ensheathed by aponeuroses (RECTAL SHEATH) of the lateral muscles of the abdominal wall
  - Segmented by 3 **tendinous inscriptions** (bands of fibrous tissue)
  - Extend medially to <u>insert</u> on the **LINEA ALBA** ("white line"), which is a tendinous raphe/ seam that extends from sternum to pubic symphysis
  O – pubic crest and symphysis
  I – xiphoid process and costal cartilages of ribs 5-7
  A – FLEX and ROTATE LUMBAR REGION of VERTEBRAL COLUMN; fix and depress ribs, stabilize pelvis during walking, increase intra-abdominal pressure; used in sit-ups/curls

LATERAL ABDOMINAL WALLS are comprised of 3 layers of broad, flat muscle sheets (SUPERFICIAL to DEEP):

- **EXTERNAL OBLIQUE:** largest of the three layers; most superficial layer; fibers run inferiorly and medially *("hands in pockets" orientation)*; its aponeurosis turns under inferiorly, forming the **inguinal ligament**

  O – by fleshy strips from outer surfaces of lower eight ribs

  I – most fibers insert into linea alba via a broad aponeurosis

  A – FLEX VERTEBRAL COLUMN and COMPRESS ABDOMINAL WALL when pair contracts simultaneously; acting individually, AID muscles of back in TRUNK ROTATION and LATERAL FLEXION; used in oblique curls

- **INTERNAL OBLIQUE:** most fibers run superiorly and medially (opposite that of external oblique muscles)

  O – lumbar fascia, iliac crest, and inguinal ligament

  I – linea alba, pubic crest, last three or four ribs and costal margin

  A – SAME AS FOR EXTERNAL OBLIQUE

- **TRANSVERSUS ABDOMINIS:** deepest (innermost) muscle layer of abdominal wall; fibers run horizontally

  O – inguinal ligament, lumbar fascia, cartilages of last six ribs; iliac crest

  I – linea alba, pubic crest

  A – COMPRESSES ABDOMINAL CONTENTS

Chapter 11 – THE MUSCULAR SYSTEM III:
MUSCLES of the APPENDICULAR SKELETON

PART 1 : MUSCLES OF THE PECTORAL GIRDLE and THE UPPER LIMB or EXTREMITY
(**O** = ORIGIN / **I** = INSERTION / **A** = ACTION)

A. **SUPERFICIAL MUSCLES of the ANTERIOR and POSTERIOR THORAX** :
MOVEMENTS of the SCAPULA

MUSCLES of the ANTERIOR THORAX:

- **PECTORALIS MINOR**: flat, thin muscle directly beneath and obscured by pectoralis major
  O – anterior surfaces of ribs 3-5 (or 2-4)
  I – coracoid process of scapula
  A - DRAWS SCAPULA FORWARD and DOWNWARD, <u>with ribs fixed</u>; draws rib cage superiorly, with scapula fixed

- **SERRATUS ANTERIOR**: fan-shaped muscle; lies deep to scapula, deep and inferior to pectoral muscles on lateral rib cage; forms medial wall of axilla; <u>origins have serrated, or sawtooth, appearance</u>
  O – by a series of muscle slips from ribs 1-8 (or 9)
  I – entire anterior surface of vertebral border of scapula
  A – ROTATES SCAPULA so that ITS INFERIOR ANGLE MOVES LATERALLY and UPWARD (upward rotation); prime mover to protract and hold scapula against chest wall

- **SUBCLAVIUS**: small cylindrical muscle extending from rib 1 to clavicle
  O – costal cartilage of rib 1
  I – groove on inferior surface of clavicle
  A – HELPS STABILIZE and DEPRESS PECTORAL GIRDLE

MUSCLES of the POSTERIOR THORAX:

- **TRAPEZIUS**: most superficial muscle of posterior thorax; triangular in shape; upper fibers run inferiorly, middle fibers run horizontally and lower fibers run superiorly to scapula
  O – occipital bone, ligamentum nuchae, and spines of C_7 and T_1 - T_{12}
  I – clavicle and scapula (acromion and scapular spine)
  A - STABILIZES, RAISES, RETRACTS, DEPRESS and ROTATES SCAPULA and/or clavicle, depending on active region and state of other muscles; can also extend neck

- **LEVATOR SCAPULAE:** located at back and side of neck, deep to trapezius; thick, straplike muscle
  O – transverse processes of C_1 - C_4
  I – medial border of the scapula, superior to the spine
  A –ELEVATES/ADDUCTS SCAPULA in concert with superior fibers of trapezius

- **RHOMBOID MAJOR and MINOR:** two rectangular muscles lying deep to trapezius
  O – spinous processes of C_7 and T_1 (<u>minor</u>) and spinous processes of T_2 - T_5 (<u>major</u>)
  I – medial border of scapula
  A – STABILIZE SCAPULA; adduct and perform downward rotation of the scapula

B. | **MUSCLES CROSSING the SHOULDER JOINT** | : MOVEMENTS OF THE HUMERUS

- **PECTORALIS MAJOR:** large, fan-shaped muscle covering upper portion of chest; divided into clavicular and sternal parts; forms anterior axillary fold
  O – sternal and inferior portion of clavicle, body of sternum, cartilage of ribs 1-6 (or 7), and aponeurosis of external oblique muscle
  I – fibers converge to insert by a short tendon into greater tubercle and lateral lip of intertubercular sulcus of humerus
  A - PRIME MOVER of ARM FLEXION; ROTATES ARM MEDIALLY; ADDUCTS ARM against resistance

- **LATISSIMUS DORSI:** broad, flat, triangular muscle of lumbar region (lower back); covered by trapezius superiorly; contributes to posterior wall of axilla
  O – spinous processes of inferior thoracic and all lumbar vertebrae, ribs 8-12, and thoracolumbar fascia
  I – spirals around teres major to insert into floor of intertubercular sulcus of the humerus
  A - PRIME MOVER of ARM EXTENSION; powerful ARM ADDUCTOR; MEDIALLY ROTATES ARM at SHOULDER; depresses scapula

- **DELTOID:** thick, multipennate muscle forming rounded shoulder muscle mass; site commonly used for intramuscular (*i.m.*) injection
  O – clavicle and scapula (acromion and adjacent scapular spine)
  I – deltoid tuberosity of humerus
  A - PRIME MOVER of ARM ABDuction(*) <u>when all its fibers contract simultaneously</u>; (anterior part) – flexion and medial rotation of humerus; (posterior part) – extension and lateral rotation of humerus

> (*) **The purpose of the typing variations "ABDuction" and "ABDuct" is to emphasize its distinction from "A<u>D</u>DUCTION" and "A<u>D</u>DUCT," respectively, making it easier to read and remember.**

- **TERES MAJOR** (anterior muscle): flat, thin muscle directly beneath and obscured by pectoralis major
  O – inferior angle of scapula
  I – medial lip of intertubercular sulcus of humerus
  A – EXTENDS, MEDIALLY ROTATES, and ADDUCTS HUMERUS

- **SUBSCAPULARIS:** forms part of posterior wall of axilla; a rotator cuff muscle
  O – subscapular fossa of scapula
  I – lesser tubercle of humerus
  A – CHIEF MEDIAL ROTATOR of HUMERUS; helps to hold head of humerus in glenoid cavity, thereby stabilizing shoulder joint

- **SUPRASPINATUS:** named for its location on posterior aspect of scapula; deep to trapezius; a rotator cuff muscle
  O – supraspinous fossa of scapula
  I – superior part of greater tubercle of humerus
  A – INITIATES ABDuction at shoulder; stabilizes shoulder joint

- **INFRASPINATUS:** partially covered by deltoid and trapezius; named for its scapular location; a rotator cuff muscle
  O – infraspinous fossa of scapula
  I – greater tubercle of humerus posterior to insertion of supraspinatus
  A – ROTATES HUMERUS LATERALLY; helps to hold head of humerus in glenoid cavity, thereby stabilizing shoulder joint

- **TERES MINOR:** small, elongated muscle located inferior to infraspinatus; a rotator cuff muscle
  O – lateral border of dorsal scapular surface
  I – greater tubercle of humerus inferior to infraspinatus insertion
  A – ROTATES HUMERUS LATERALLY; helps to hold head of humerus in glenoid cavity, thereby stabilizing shoulder joint; adduction at shoulder

- **CORACOBRACHIALIS:** small, cylindrical muscle
  O – coracoid process of scapula
  I – medial surface of humerus shaft
  A – FLEXION and ADDUCTION of the HUMERUS; synergist of pectoralis major

C. ⬚ **MUSCLES CROSSING the ELBOW JOINT** ⬚ : **FLEXION** and **EXTENSION** of the **FOREARM**

(FLEXORS are found on ANTERIOR aspect; EXTENSORS on POSTERIOR aspect)

POSTERIOR MUSCLES crossing the elbow joint

- **TRICEPS BRACHII**: large fleshy muscle, solely occupying the posterior compartment of the arm; three-headed origin; long and lateral heads lie superficial to medial head
  O – <u>long head</u>: infraglenoid tubercle of scapula; <u>lateral head</u>: posterior shaft of humerus; <u>medial head</u>: posterior humeral shaft distal to radial groove
  I – by common tendon into olecranon process of ulna
  A - POWERFUL FOREARM EXTENSOR (PRIME MOVER = MEDIAL HEAD); antagonist of forearm flexors; (<u>long head</u>): extension and adduction at shoulder also

- **ANCONEUS**: short triangular muscle; closely associated (blended) with distal end of triceps on posterior humerus
  O – lateral epicondyle of humerus
  I – lateral aspect of olecranon process of ulna
  A – ABDucts ULNA <u>during forearm PRONATION</u>; synergist of triceps brachii in elbow extension

ANTERIOR MUSCLES crossing the elbow joint

- **BICEPS BRACHII**: two-headed fusiform muscle; bellies unite as insertion point is approached; tendon of long head helps stabilize shoulder joint
  O – <u>short head</u>: coracoid process of scapula; <u>long head</u>: supraglenoid tubercle and lip of glenoid cavity of scapula
  I – by common tendon into radial tuberosity
  A - FLEXION at ELBOW and SHOULDER and SUPINATES FOREARM, usually simultaneously

- **BRACHIALIS**: strong muscle immediately deep to biceps brachii on distal humerus
  O – anterior, distal surface of humerus
  I – coronoid process of ulna and capsule of elbow joint
  A - A MAJOR FOREARM FLEXOR (lifts ulna as biceps lifts the radius)

- **BRACHIORADIALIS**: superficial muscle of lateral forearm, extending from distal humerus to distal forearm
  O – lateral supracondylar ridge at distal end of humerus
  I – base of styloid process of radius
  A - SYNERGIST in FOREARM FLEXION; stabilizes the elbow during rapid flexion <u>and</u> extension

D. | Muscles of the FOREARM | : MOVEMENTS of the WRIST, HAND, and FINGERS

(FLEXORS are found on the <u>anterior</u> aspect; EXTENSORS on the <u>posterior</u> aspect)

| TENDONS and MUSCLES of the HAND | :

- **FLEXOR and EXTENSOR RETINACULA** – band-like thickenings of deep fascia, which firmly anchor the long tendons (of the forearm muscles) that insert in the hand distally. The FLEXOR RETINACULUM and the EXTENSOR RETINACULUM keep the tendons from jumping outward when the hand is hyperflexed or hyperextended, respectively.

- **PALMAR APONEUROSIS** – sheet-like tendinous extension of the insertion tendon of the palmaris longus muscle

---

ANTERIOR MUSCLES of the FOREARM: Most of the tendons of insertion of these FLEXORS are held in place by the FLEXOR RETINACULUM. The median nerve or its branches and branches of the ulnar nerve innervate all these muscles.

| SUPERFICIAL MUSCLES of the ANTERIOR FOREARM |

- **PRONATOR TERES:** two-headed muscle; seen in superficial view between proximal margins of brachioradialis and flexor carpi radialis
  O – medial epicondyle of humerus; coronoid process of ulna
  I – by common tendon into lateral radius, midshaft
  A - PRONATES FOREARM; weak flexor of elbow

- **FLEXOR CARPI RADIALIS:** runs diagonally across forearm; midway, its fleshy belly is replaced by a flat tendon that becomes cord-like at wrist
  O – medial epicondyle of humerus
  I – bases of 2nd and 3rd metacarpal bones
  A - POWERFUL FLEXOR of WRIST; ABDucts HAND; weak synergist of elbow flexion

- **PALMARIS LONGUS:** small, fleshy muscle with a long insertion tendon, which is located superficial to the flexor retinaculum, making this a **landmark anterior muscle**; its tendon inserts into the palm as the **PALMAR APONEUROSIS**
  O – medial epicondyle of humerus
  I – fascia of palm (palmar aponeurosis) and flexor retinaculum
  A - TENSES SKIN and FASCIA OF PALM during HAND MOVEMENTS; weak wrist flexor; weak synergist for elbow flexion

○ **FLEXOR CARPI ULNARIS**: most medial anterior muscle; two-headed; ulnar nerve lies lateral to its tendon

O – medial epicondyle of humerus; olecranon process and posterior surface of ulna

I – pisiform and hamate bones and base of 5th metacarpal

A - POWERFUL FLEXOR of WRIST; ADDUCTS HAND in concert with extensor carpi ulnaris (posterior muscle);

○ **FLEXOR DIGITORUM SUPERFICIALIS**: two-headed muscle; more deeply placed; overlain by muscles above but visible at distal end of forearm

O – medial epicondyle of humerus, coronoid process of ulna; shaft of radius

I – by four tendons into middle phalanges of fingers 2-5

A – FLEXION at MIDDLE PHALANGES of FINGERS 2-5 (proximal interphalangeal and metacarpophalangeal joints) and WRIST JOINTS

DEEP MUSCLES of the ANTERIOR FOREARM

○ **FLEXOR POLLICIS LONGUS**: partly covered by flexor digitorum superficialis; lies lateral and parallel to flexor digitorum profundus

O – anterior surface of radius and interosseous membrane

I – distal phalanx of thumb

A – FLEXES DISTAL PHALANX of THUMB

○ **FLEXOR DIGITORUM PROFUNDUS** (deep): extensive origin; overlain entirely by flexor digitorum superficialis

O – anteromedial surface of ulna and interosseous membrane

I – by four tendons into distal phalanges of fingers 2-5

A – FLEXES DISTAL INTERPHALANGEAL JOINTS; slow-acting flexor of any or all fingers; assists in flexing wrist

○ **PRONATOR QUADRATUS**: deepest muscle of distal forearm; passes downward and laterally; only muscle that arises solely from ulna and inserts solely into radius

O – distal portion of anterior ulnar shaft

I – distal surface of anterior radius

A – PRIME MOVER of FOREARM PRONATION; acts with pronator teres; also helps hold ulna and radius together

---

POSTERIOR MUSCLES of the FOREARM: The tendons of the EXTENSORS are held in place at the posterior aspect of the wrist by the EXTENSOR RETINACULUM, which prevents "bowstringing" of these tendons when the wrist is hyperextended. The radial nerve or its branches innervate all these muscles.

SUPERFICIAL MUSCLES of the POSTERIOR FOREARM

o **BRACHIORADIALIS:** This most-anterior muscle of the posterior fascial compartment is visible on the anterior forearm. It is also categorized in the muscle group crossing the elbow joint due to its origin in the distal humerus.
O - See Section C – Anterior Muscles Crossing the Elbow Joint
I - See Section C – Anterior Muscles Crossing the Elbow Joint
A - See Section C – Anterior Muscles Crossing the Elbow Joint

o **EXTENSOR CARPI RADIALIS LONGUS:** parallels brachioradialis on lateral forearm, and may blend with it
O – lateral supracondylar ridge of humerus
I – base of second metacarpal
A - EXTENDS WRIST <u>in concert with the extensor carpi ulnaris</u>; and ABDucts WRIST <u>in concert with the flexor carpi radialis</u>

o **EXTENSOR CARPI RADIALIS BREVIS:** somewhat shorter than extensor carpi radialis longus and lies deep to it
O – lateral epicondyle of humerus
I – base of 3rd metacarpal
A – EXTENDS and ABDucts WRIST; acts synergistically with extensor carpi radialis longus to stabilize wrist during finger flexion

o **EXTENSOR DIGITORUM:** lies medial to extensor carpi radialis brevis
O – lateral epicondyle of humerus
I – by 4 tendons into extensor expansions and distal phalanges of digits 2-5
A - a detached portion of this muscle (EXTENSOR DIGITI MINIMI), extends the little finger; PRIME MOVER of FINGER EXTENSION; extends wrist; can ABDuct (flare) fingers

o **EXTENSOR CARPI ULNARIS:** most medial of superficial posterior muscles; long, slender muscle
O – lateral epicondyle of humerus and posterior border of ulna
I – base of 5th metacarpal
A - EXTENDS WRIST <u>in concert with the extensor carpi radialis</u>; and ADDUCTS WRIST <u>in concert with the flexor carpi ulnaris</u>

DEEP MUSCLES of the POSTERIOR FOREARM

o **SUPINATOR:** deep muscle at posterior aspect of elbow; largely concealed by superficial muscles
O – lateral epicondyle of humerus; proximal ulna
I – proximal end of radius

125

A – ASSISTS BICEPS BRACHII to FORCIBLY SUPINATE FOREARM; works alone in slow supination; antagonist of pronator muscles

o **ABDuctor POLLICIS LONGUS:** lateral and parallel to extensor pollicis longus; just distal to supinator
 O – posterior surface of radius and ulna; interosseous membrane
 I – base of 1st metacarpal and trapezium
 A – ABDucts and EXTENDS THUMB (pollex); ABDuction at wrist joint

o **EXTENSOR POLLICIS BREVIS and LONGUS:** deep muscle pair with a common origin and action; overlain by the extensor carpi ulnaris
 O – dorsal shaft of radius and ulna; interosseous membrane
 I – base of proximal (brevis) and distal (longus) phalanx of thumb
 A - EXTENDS the THUMB (pollex); ABDuction at wrist joint

o **EXTENSOR INDICIS:** tiny muscle arising close to wrist
 O – posterior surface of distal ulna; interosseous membrane
 I – extensor expansion of index finger; joints tendon of extensor digitorum
 A – EXTENDS INDEX FINGER and assists in wrist extension

E. | **INTRINSIC MUSCLES of the HAND** |: **FINE MOVEMENTS of the FINGERS**

This group of muscles considers the small muscles that lie entirely in the hand. All are in the palm, none in the dorsum of the hand. All these muscles are small and weak; they mostly control precise movements of the metacarpals and fingers, as in threading a needle.

The intrinsic muscles of the palm are divided into three groups: 1) those in the THENAR EMINENCE (ball of the thumb); 2) those in the HYPOTHENAR EMINENCE (ball of the little finger; 3) muscles in the MID-PALM

THENAR MUSCLES in the BALL of the THUMB (*thenar* = palm; *pollex* = thumb):

• **ABDuctor POLLICIS BREVIS** (superficial): lateral muscle of thenar group
 O – flexor retinaculum and nearby carpals
 I – lateral base of thumb's proximal phalanx
 A – ABDucts THUMB at carpometacarpal (*"ball of thumb"*) joint

- **FLEXOR POLLICIS BREVIS** (deep): medial and deep muscle of thenar group
  O – flexor retinaculum and nearby carpals
  I – lateral side of base of proximal phalanx of thumb
  A - FLEXES and ADDUCTS THUMB at <u>carpometacarpal</u> and <u>metacarpophalangeal</u> (*"knuckle"*) joints

- **OPPONENS POLLICIS** (deep): deep to abductor pollicis brevis, on metacarpal 1
  O – flexor retinaculum and trapezium
  I – whole anterior side of metacarpal 1
  A – OPPOSITION of THUMB; moves thumb to touch tip of little finger

- **ADDUCTOR POLLICIS** (deep): fan-shaped with horizontal fibers; distal to other thenar muscles; oblique and transverse heads
  O – capitate bone and bases of metacarpals 2 - 4; front of metacarpal 3
  I – medial side of base of proximal phalanx of thumb
  A - ADDUCTS and HELPS to OPPOSE the THUMB

HYPOTHENAR MUSCLES in the BALL of the LITTLE FINGER

- **ABDuctor DIGITI MINIMI:** medial muscle of hypothenar; <u>superficial</u>
  O – pisiform bone
  I – medial side of proximal phalanx of little finger
  A – ABDucts LITTLE FINGER at <u>metacarpophalangeal</u> joint

- **FLEXOR DIGITI MINIMI BREVIS:** lateral <u>deep</u> muscle of hypothenar group
  O – hamate bone and flexor retinaculum
  I – medial side of proximal phalanx of little finger
  A – FLEXES LITTLE FINGER at <u>metacarpophalangeal</u> joint

- **OPPONENS DIGITI MINIMI:** <u>deep</u> to abductor digiti minimi
  O – hamate bone and flexor retinaculum
  I – most of length of medial side of metacarpal 5
  A – HELPS in OPPOSITION; brings metacarpal 5 toward thumb to cup the hand; flexion at metacarpophalangeal joint

---

MIDPALMAR MUSCLES

- **LUMBRICALS:** worm-shaped muscles in palm, one to each finger (except thumb); unusual because they originate from tendons of another muscle
  O – lateral side of each tendon of flexor digitorum profundus in palm
  I – lateral edge of extensor expansion on first phalanx of fingers 2-5
  A – FLEXION at METACARPOPHALANGEAL JOINT; EXTENSION at PROXIMAL and DISTAL INTERPHALANGEAL JOINTS

- **PALMAR INTEROSSEI** (singular = *interosseus*): <u>four</u> long, cone-shaped muscles in the spaces between the metacarpals; lie ventral to the dorsal interossei
  O – sides of metacarpal bones 2, 4, and 5
  I – bases of proximal phalanges of digits 2,4, and 5
  A – ADDUCTION of FINGERS at METACARPOPHALANGEAL JOINTS of DIGITS 2, 4, and 5; flexion at metacarpophalangeal joints; extension at interphalangeal joints

- **DORSAL INTEROSSEI** (singular = *interosseus*): <u>four</u> bipennate muscles filling spaces between the metacarpals; deepest palm muscles, also visible on dorsal side of hand
  O – each originates from opposing faces of two metacarpal bones (1 and 2, 2 and 3, 3 and 4, 4 and 5)
  I – bases of proximal phalanges of digits 2-4
  A – ABDuct (DIVERGE) FINGERS at METACARPOPHALANGEAL JOINTS of DIGITS 2-4; flexion at metacarpophalangeal joints; extension at interphalangeal joints

---

PART 2 : MUSCLES OF THE PELVIC DIAPHRAGM and the LOWER LIMB or EXTREMITY

F. **PELVIC DIAPHRAGM**

- **LEVATOR ANI:** broad, thin, tripartite muscle (ILIOCOCCYGEUS and PUBOCOCCYGEUS) whose fibers extend inferomedially (inferior and medial), forming a muscular "sling" around the prostate(male)/vagina(female), urethra and anorectal junction before meeting in the median plane
  O – extensive linear origin inside pelvis from the pubis to the ischial spine of coxal bone
  I – inner surface of coccyx, levator ani of opposite side, and partially into structures that penetrate it
  A – SUPPORTS and MAINTAINS POSITION of PELVIC VISCERA; resists downward thrusts that accompany rises in intrapelvic pressure during coughing, vomiting, and expulsive efforts of abdominal muscles; lifts anal canal during defecation

    - **ILIOCOCCYGEUS: horizontal fibers of levator ani**
      O – ischial spine, pubis
      I – coccyx and median raphe
      A – tenses floor of pelvis, supports pelvic organs, flexes coccygeal joints, elevates and retracts anus

    - **PUBOCOCCYGEUS: vertical fibers of levator ani**
      O – inner margins of pubis
      I – coccyx and median raphe
      A – SAME AS FOR ILIOCOCCYGEUS

- **COCCYGEUS:** small, triangular muscle lying posterior to levator ani; posterior part of pelvic diaphragm
  O – spine of ischium

I – sacrum and coccyx
A – SUPPORTS PELVIC VISCERA and coccyx; pulls coccyx forward after it has been reflected posteriorly by defecation and childbirth

- **EXTERNAL ANAL SPHINCTER:** flat plane of muscular fibers, elliptical in shape and intimately adherent to the integument surrounding the margin of the anus
  O – via tendon from coccyx
  I – encircles anal opening
  A – CLOSES the ANAL OPENING (or anal orifice)

G. ☐ **UROGENITAL DIAPHRAGM** ☐

- **DEEP TRANSVERSE PERINEAL MUSCLE:** together, the pair spans distance between ischial rami; in females, lies posterior to vagina
  O – ischial rami
  I – median raphe of urogenital diaphragm (midline central tendon of perineum); some fibers insert into vaginal wall in females
  A – SUPPORTS PELVIC ORGANS; stabilizes central tendon

- **EXTERNAL URETHRAL SPHINCTER:** muscle encircling the urethra (in both males and females) and the vagina (in females)
  O – ischiopubic rami
  I – midline raphe
  A – CONSTRICTS URETHRA; allows voluntary inhibition of urination; helps support pelvic organs

H. ☐ **MUSCLES OF THE SUPERFICIAL PERINEAL SPACE** ☐

- **ISCHIOCAVERNOSUS:** runs from pelvis to base of penis or clitoris
  O – ischial tuberosities
  I – crus of corpus cavernosa of penis or clitoris
  A - RETARDS VENOUS DRAINAGE and MAINTAINS ERECTION of PENIS or CLITORIS

- **BULBOSPONGIOSUS:** encloses base of penis (bulb) in males and lies deep to labia in females
  O – in males and females, the perineal body (central tendon of perineum); and in males, the median raphe of the penis also
  I – anteriorly into corpus cavernosa of penis or clitoris; corpus spongiosum and perineal membrane (males); bulb of vestibule, perineal membrane , and body of clitoris (females)
  A - EMPTIES MALE URETHRA; assists in erection of penis and clitoris – compresses base, stiffens penis / compresses and stiffens clitoris; narrows vaginal opening

- **SUPERFICIAL TRANSVERSE PERINEAL MUSCLE:** paired muscle bands posterior to urethral (and in females, vaginal) opening; variable; sometimes absent
  O – ischial tuberosity
  I – central tendon of perineum
  A - STABILIZES and STRENGTHENS CENTRAL TENDON of PERINEUM

I. | ANTERIOR and MEDIAL MUSCLES of the THIGH | : ORIGIN on the PELVIS or BACKBONE

- **ILIOPSOAS:** composite of 2 closely related muscles whose fibers pass under the inguinal ligament; both muscles insert via a common tendon on the femur (ILIOPSOAS TENDON)

  - **ILIACUS:** large, fan-shaped; more lateral
    O – iliac fossa, ala of sacrum
    I – lesser trochanter of femur via iliopsoas tendon
    A – PRIME MOVER in THIGH FLEXION and FLEXING the TRUNK

  - **PSOAS MAJOR:** longer, thicker; more medial
    O – by fleshy slips from transverse processes, bodies and discs of $L_1 - L_5$ and T_{12}
    I – lesser trochanter of femur via iliopsoas tendon
    A – same actions as that of ILIACUS; also LATERAL FLEXION of VERTEBRAL COLUMN; postural muscle

- **SARTORIUS:** longest muscle in the body; strap-like; superficial; runs obliquely across anterior surface of thigh to knee; crosses both hip and knee joints
  O – anterior superior iliac spine
  I – winds around medial aspect of knee and inserts into medial aspect of proximal tibia
  A – FLEXES, ABDucts and LATERALLY ROTATES THIGH; FLEXES KNEE

J. | MEDIAL COMPARTMENT of the THIGH |

MEDIAL MUSCLES: ADDUCTOR muscles; move the thigh only

- | ADDUCTORS | **(longus, brevis, and magnus):** this large muscle mass consists of three muscles forming the medial aspect of the thigh; arise from inferior part of pelvis and insert at various levels on femur; innervated by OBTURATOR NERVE; strain or stretching of this muscle group is called a "PULLED GROIN"

  - **ADDUCTOR LONGUS:** overlies middle aspect of adductor magnus; most anterior of adductor muscles
    O - pubis near pubic symphysis
    I – linea aspera of femur
    A - ADDUCTS, FLEXES, and MEDIALLY ROTATES THIGH

- **ADDUCTOR BREVIS:** in contact with obturator externus muscle; largely concealed by adductor longus and pectineus
  O – body and inferior ramus of pubis
  I – linea aspera above adductor longus
  A - ADDUCTS and MEDIALLY ROTATES THIGH

- **ADDUCTOR MAGNUS:** a triangular muscle with a broad insertion; is a composite muscle that is part adductor and part hamstring in action
  O – ischial and pubic rami and ischial tuberosity
  I – linea aspera and adductor tubercle of femur
  A - whole muscle produces ADDUCTION at the HIP; <u>anterior part</u> MEDIALLY ROTATES and FLEXES THIGH; <u>posterior part</u> is a synergist of hamstrings in THIGH EXTENSION

- **PECTINEUS:** short, flat muscle; overlies adductor brevis on proximal thigh; abuts adductor longus medially
  O – pectineal line of pubis (and superior ramus)
  I – a line from lesser trochanter to linea aspera on posterior aspect of femur
  A – ADDUCTS, FLEXES and MEDIALLY ROTATES THIGH

- **GRACILIS:** long, thin, superficial muscle of medial thigh
  O – inferior ramus and body of pubis and adjacent ischial ramus
  I – medial surface of tibia just inferior to its medial condyle
  A – ADDUCTS and MEDIALLY ROTATES THIGH; FLEXES and MEDIALLY ROTATES LEG, especially during walking

K. | **ANTERIOR COMPARTMENT of the THIGH** |

ANTERIOR MUSCLES: **flex the femur at the hip and extend leg at the knee (foreswing phase of walking)**

- | **QUADRICEPS FEMORIS** | : flesh of front and sides of the thigh; quadriceps tendon inserts into patella, then continues into tibial tuberosity via patellar ligament

  - **POWERFUL KNEE EXTENSOR**

  - ARISES from FOUR HEADS:

    - **RECTUS FEMORIS:** superficial muscle of anterior thigh; runs straight down thigh
      O – anterior inferior iliac spine and superior acetabular rim of ilium
      I – tibial tuberosity via quadriceps tendon, patella, and patellar ligament
      A – EXTENDS KNEE and FLEXES THIGH at HIP

- **VASTUS LATERALIS** (lateral):   largest head of the group, forms lateral aspect of thigh
  O – greater trochanter, intertrochanteric line, linea aspera of femur
  I – as for rectus femoris
  A – EXTENDS and STABILIZES KNEE

- **VASTUS MEDIALIS** (inferomedial):
  O – linea aspera, medial supracondylar line, intertrochanteric line
  I – as for rectus femoris
  A – EXTENDS KNEE; inferior fibers stabilize patella

- **VASTUS INTERMEDIUS** (obscured by rectus femoris):
  O – anterior and lateral surfaces of proximal femur shaft
  I – as for rectus femoris
  A – EXTENDS KNEE

- | **TENSOR FASCIAE LATAE** | : **muscle** that is enclosed between fascia layers of anterolateral aspect of thigh; functionally associated with medial rotators and flexors of thigh
  O – anterior aspect of iliac crest and anterior superior iliac spine
  I – iliotibial tract (thickened lateral portion of the *fascia lata**)
  A – STEADIES TRUNK on THIGH by MAKING ILIOTIBIAL TRACT TAUT; flexes and ABDucts thigh; rotates thigh medially; extension and lateral rotation at knee

  > ***FASCIA LATA:**   deep **fascia** of the thigh that surrounds, encloses, and separates the anterior, medial and posterior muscle compartments, like a support stocking

L. | **POSTERIOR COMPARTMENT of the THIGH** |

POSTERIOR MUSCLES: **mostly extend thigh and flex leg (backswing phase of walking)**

| **GLUTEAL MUSCLES** | **– ORIGIN ON PELVIS OR SACRUM:**

- **GLUTEUS MAXIMUS:** the largest and most superficial of gluteal muscles; forms bulk of buttock mass
  O – dorsal ilium, sacrum and coccyx
  I – gluteal tuberosity of femur; iliotibial tract
  A – MAJOR EXTENSOR of THIGH; lateral rotation of thigh; helps stabilize the extended knee; ABDuction at the hip (superior fibers only)

- **GLUTEUS MEDIUS:** thick muscle largely covered by gluteus maximus; important site for intramuscular injections (ventral gluteal site)
  O –between anterior and posterior gluteal lines on lateral surface of ilium
  I – by short tendon into lateral aspect of greater trochanter of femur
  A – ABDucts and MEDIALLY ROTATES THIGH

- **GLUTEUS MINIMUS:** smallest and deepest of gluteal muscles
  O – lateral surface of ilium between inferior and anterior gluteal lines
  I – anterior border of greater trochanter of femur
  A – ABDucts and MEDIALLY ROTATES THIGH

LATERAL ROTATORS

- **PIRIFORMIS:** pyramidal muscle located on posterior aspect of hip joint; inferior to gluteus minimus
  O – anterolateral surface of sacrum (opposite greater sciatic notch)
  I – superior border of greater trochanter of femur
  A - ROTATES EXTENDED THIGH LATERALLY; ABDuction of hip; help to maintain stability and integrity of the hip

- **OBTURATOR EXTERNUS and INTERNUS:** flat, triangular muscle deep in upper medial aspect of thigh
  O – lateral and media margins of obturator foramen
  I – trochanteric fossa of femur (externus); medial surface of greater trochanter (internus)
  A - SAME ACTIONS AS PIRIFORMIS

- **GEMELLUS SUPERIOR and INFERIOR** (plural = *gemelli*)**:** two small muscles with common insertions and actions; considered extrapelvic portions of obturator internus
  O – ischial spine (<u>superior</u>); ischial tuberosity (<u>inferior</u>)
  I – greater trochanter of femur
  A - SAME ACTIONS AS PIRIFORMIS

- **QUADRATUS FEMORIS:** short, thick muscle; most inferior of lateral rotator muscles; extends laterally from pelvis
  O – ischial tuberosity
  I – intertrochanteric crest of femur
  A - ROTATES THIGH LATERALLY and STABILIZES HIP JOINT

MUSCLES OF THE POSTERIOR COMPARTMENT OF THE THIGH :

HAMSTRINGS: a group of three fleshy muscles of the posterior thigh / cross both hip and knee joints / **PRIME MOVERS of THIGH EXTENSION and KNEE FLEXION**

- **BICEPS FEMORIS:** most **lateral** muscle of the group; arises from 2 heads
  O – ischial tuberosity (<u>long head</u>), linea aspera, lateral supracondylar line and distal femur (<u>short head</u>)
  I – common tendon passes downward and laterally (forming lateral border of popliteal fossa) to insert into head of fibula and lateral condyle of tibia
  A – EXTENDS THIGH and FLEXES KNEE; laterally rotates leg, especially when knee is fixed

- **SEMITENDINOSUS:** **medial** to biceps femoris
  O – ischial tuberosity
  I – medial aspect of upper tibial shaft
  A – EXTENDS THIGH at HIP; FLEXES KNEE; with semimembranosus, medially rotates leg

- **SEMIMEMBRANOSUS:** **deep** to semitendinosus
  O – ischial tuberosity
  I – medial condyle of tibia; through oblique ligament to lateral condyle of femur
  A – EXTENDS THIGH and FLEXES KNEE; medially rotates leg as above

M. ANTERIOR COMPARTMENT of the LEG

*The crural (lower leg) muscles are arranged opposite to that in the antebrachium (forearm) muscles: ANTERIOR COMPARTMENT muscles are EXTENSORS whereas POSTERIOR COMPARTMENT muscles are FLEXORS.

EXTENSOR MUSCLES : extend the toes; dorsiflex the foot at the ankle

- **TIBIALIS ANTERIOR:** superficial; laterally parallels sharp anterior margin of tibia
  O – lateral condyle and upper 2/3 of tibial shaft
  I – by tendon into inferior surface of medial cuneiform and 1st metatarsal
  A – PRIME MOVER of DORSIFLEXION; inverts foot

- **EXTENSOR DIGITORUM LONGUS:** unipennate muscle (anterolateral surface of leg) lateral to tibialis anterior
  O – lateral condyle of tibia; proximal ¾ of fibula
  I – middle and distal phalanges of toes/digits 2-5
  A – PRIME MOVER of TOE (2-5) EXTENSION; dorsiflexes foot (at ankle)

- **FIBULARIS (PERONEUS) TERTIUS:** small muscle; usually continuous and fused with distal part of extensor digitorum longus; **not always present**
  O – distal anterior surface of fibula and interosseous membrane
  I – tendon inserts on dorsum of 5th metatarsal
  A - DORSIFLEXES and EVERTS FOOT

- **EXTENSOR HALLUCIS LONGUS:** deep, narrow origin
  O – anteromedial fibula shaft
  I – tendon inserts on distal phalanx of great toe
  A – EXTENDS GREAT TOE; dorsiflexes at ankle (foot)

N. | **LATERAL COMPARTMENT of the LEG** |

PLANTAR FLEXION / FOOT EVERSION / STABILIZE LATERAL ANKLE and LATERAL LONGITUDINAL ARCH of the FOOT

- **FIBULARIS (PERONEUS) LONGUS:** superficial lateral muscle; overlies fibula
  O – head and upper portion of lateral side of fibula
  I – by long tendon that curves under foot to 1st metatarsal and medial cuneiform
  A – PLANTAR FLEXES and EVERTS FOOT; supports ankle; supports longitudinal and transverse arches

- **FIBULARIS (PERONEUS) BREVIS:** smaller muscle; deep to fibularis longus; enclosed in a common sheath
  O – distal fibula shaft
  I – by tendon running behind lateral malleolus to insert on proximal end of the 5th metatarsal
  A – PLANTAR FLEXES at ANKLE and EVERTS FOOT

O. | **POSTERIOR COMPARTMENT of the LEG** (SURAL REGION) |

PLANTAR FLEXION of the ANKLE

SUPERFICIAL MUSCLES of the SURAL REGION

- **TRICEPS SURAE:** muscle pair that shapes posterior calf and inserts via a common tendon into calcaneus = **CALCANEAL or ACHILLES TENDON** (largest tendon in the body)
  PRIME MOVERS of ANKLE PLANTAR FLEXION

  - **GASTROCNEMIUS: superficial** muscle of the pair – two prominent bellies form the proximal curve of calf
    O – by 2 heads from medial and lateral condyles of femur
    I – posterior calcaneus via calcaneal tendon
    A – PLANTAR FLEXES FOOT <u>when knee is extended</u>; flexion at knee

  - **SOLEUS:** broad, flat muscle; **deep** to gastrocnemius; on posterior surface of calf
    O – extensive cone-shaped origin from superior tibia, fibula and interosseous membrane
    I – as for gastrocnemius
    A – PLANTAR FLEXES FOOT; postural muscle when standing

- **PLANTARIS:** generally small, feeble muscle but varies in size and extent; **may not be present**
  O – posterior femur above lateral condyle
  I – via calcaneal tendon
  A – assists in KNEE FLEXION and PLANTAR FLEXION of FOOT

DEEP MUSCLES of the SURAL REGION

- **POPLITEUS:** thin, triangular muscle at posterior knee; passes downward and medially to tibial surface
  O – lateral condyle of femur and lateral meniscus of knee
  I – proximal tibia
  A – FLEXES and ROTATES LEG MEDIALLY <u>to unlock knee from full extension when flexion begins</u>

- **FLEXOR DIGITORUM LONGUS:** long, narrow muscle / runs **medial** to and partially overlies tibialis posterior
  O – extensive origin on posterior tibia
  I – tendon runs behind the medial malleolus and splits to insert into distal phalanges of toes 2-5
  A – PLANTAR FLEXES and INVERTS FOOT; FLEXES TOES; helps foot "grip" ground

- **FLEXOR HALLUCIS LONGUS:** bipennate; lies **lateral** to inferior aspect of tibialis posterior
  O – middle part of shaft of fibula; interosseous membrane
  I – tendon runs under foot to distal phalanx of great toe
  A – PLANTAR FLEXES and INVERTS FOOT; FLEXES HALLUX (great toe) at all joints; "push off" muscle during walking

- **TIBIALIS POSTERIOR:** thick, flat muscle deep to soleus; placed between posterior flexors
  O – interosseous membrane and adjacent shafts of tibia and fibula
  I – navicular, all three cuneiforms, cuboid, 2nd, 3rd, and 4th metatarsals
  A – PRIME MOVER of FOOT INVERSION; plantar flexion at ankle

P. | **INTRINSIC MUSCLES of the FOOT** | : TOE MOVEMENT and FOOT SUPPORT

> These muscles help to **flex, extend, abduct, and adduct the toes**. These intrinsic muscles also help **support the arches of the foot**, in concert with the tendons of some leg muscles that enter the sole. On the *foot's dorsum (superior aspect)*, there is only a single muscle; while many muscles are located on the *plantar aspect (sole)* of the foot. The **plantar muscles occur in four layers** from superficial to deep. Overall, the muscles of the foot are very similar to those in the palm of the hand.

MUSCLE on the DORSUM of FOOT

- **EXTENSOR DIGITORUM BREVIS:** small, four-part muscle on dorsum of foot; deep to the tendons of extensor digitorum longus; corresponds to the extensor indicis and extensor pollicis muscles of forearm
  O – superior and lateral surfaces of calcaneus bone; extensor retinaculum
  I – base of proximal phalanx of hallux (big toe); extensor expansions on digits/toes 2-4 (dorsal surface of toes 1-4)
  A – HELPS EXTEND TOES at METATARSOPHALANGEAL JOINTS

MUSCLES on the SOLE of FOOT - four layers, *superficial to deep*

1st **LAYER** (most superficial):

- **FLEXOR DIGITORUM BREVIS:** bandlike muscle in middle of sole; corresponds to flexor digitorum superficialis of forearm and inserts into digits in the same way
  O – calcaneal tuberosity
  I – middle phalanx of digits/toes 2-5
  A – HELPS FLEX TOES 2-5 at PROXIMAL INTERPHALANGEAL joints

- **ABDuctor HALLUCIS:** lies medial to flexor digitorum brevis (analogous/corresponds to thumb muscle, abductor pollicis brevis)
  O – calcaneal tuberosity and flexor retinaculum
  I – medial side of proximal phalanx of great toe, through a tendon shared with flexor hallucis brevis
  A – ABDucts GREAT TOE (HALLUX) at <u>metatarsophalangeal joint</u>

- **ABDuctor DIGITI MINIMI:** most lateral of the three superficial sole muscles (analogous to abductor muscle in palm of hand)
  O – calcaneal tuberosity
  I – lateral side of base of digit 5's proximal phalanx
  A – ABDucts and FLEXES LITTLE TOE (DIGIT 5) at <u>metatarsophalangeal joint</u>

2nd LAYER:

- **QUADRATUS PLANTAE (FLEXOR ACCESSORIUS):** rectangular muscle just deep to flexor digitorum brevis in posterior half of sole; 2 heads
  O – medial and inferior surfaces of calcaneus
  I – tendon of flexor digitorum longus in midsole
  A – STRAIGHTENS OUT the OBLIQUE PULL of FLEXOR DIGITORUM LONGUS (i.e. FLEXION at JOINTS of TOES 2-5)

- **LUMBRICALS (4):** four little "worms" (like lumbricals in hand)
  O – from each tendon of flexor digitorum longus
  I – insertions of extensor digitorum longus (superior surfaces of phalanges, toes 2-5)
  A – FLEX TOES at METATARSOPHALANGEAL JOINTS and EXTEND TOES at INTERPHALANGEAL JOINTS, TOES 2-5

3rd LAYER:

- **FLEXOR HALLUCIS BREVIS:** covers metatarsal I: splits into two bellies (analogous to flexor pollicis brevis of thumb)
  O – lateral cuneiform and cuboid bones
  I – via two tendons onto both sides of the base of the proximal phalanx of great toe; each tendon has a sesamoid bone in it
  A – FLEXES HALLUX (big toe) at METATARSOPHALANGEAL JOINT

- **ADDUCTOR HALLUCIS:** oblique and transverse heads; deep to lumbricals (analogous to adductor pollicis in thumb)
  O – from bases of metatarsal bones II-IV and from sheath of fibularis longus tendon (oblique head); from plantar ligament across metatarsophalangeal joints (transverse head)
  I – base of proximal phalanx of great toe, lateral side
  A – HELPS MAINTAIN the TRANSVERSE ARCH of the FOOT; <u>adduction and flexion of great toe at metatarsophalangeal joint</u>

- **FLEXOR DIGITI MINIMI BREVIS:** covers metatarsal V (recall the same muscle in the hand)
  O – base of metatarsal bone V and sheath of fibularis longus tendon
  I – lateral side of proximal phalanx of toe 5
  A – FLEXES TOE 5 at METATARSOPHALANGEAL JOINT

4th LAYER:

- **PLANTAR INTEROSSEI (3):** (sing. = *interosseous*); similar to plantar interossei of hand in locations, attachments, and actions; HOWEVER, the long axis of the foot around which these muscles orient is the 2nd digit (toe), not the 3rd digit
  O – bases and medial sides of metatarsal bones
  I – medial sides of toes 3-5
  A – ADDUCTION of METATARSOPHALANGEAL JOINTS of TOES 3-5; flexion of metatarsophalangeal joints and extension at interphalangeal joints

- **DORSAL INTEROSSEI (4):** as described for plantar interossei
  O – sides of metatarsal bones
  I – medial and lateral sides of toe 2; lateral sides of toes 3 and 4
  A – ABDuction at METATARSOPHALANGEAL JOINTS of TOES 3 and 4; flexion of metatarsophalangeal joints and extension at the interphalangeal joints of toes 2-4

PRACTICE QUESTIONS 3:
The Muscular System

1. The *triceps brachii* and *rectus femoris* are good examples of:
   a. agonists.
   b. flexors.
   c. synergists.
   d. extensors.
   e. antagonists.

2. *Semitendinosus* is assisted in flexion of the thigh by the synergistic muscle:
   a. biceps femoris.
   b. pronator teres.
   c. rectus femoris.
   d. gastrocnemius.
   e. A and C

3. Which of the following muscles rotates the femur laterally?
   a. popliteus
   b. pectineus
   c. quadratus femoris
   d. obturator internus
   e. C and D

4. Which of the following muscles does _not_ belong to the group known as "hamstrings?"
   a. semimembranosus
   b. biceps brachii
   c. semitendinosus
   d. biceps femoris
   e. B and C

5. Muscle(s) in the *sural* region include:
   a. the plantaris.
   b. the soleus.
   c. the gastrocnemius
   d. A, B, and C
   e. B and C only

6. A muscle that acts as a flexor of the knee and originates on the iliac spine is the:
   a. popliteus.
   b. sartorius.

c.   pectineus.
    d.   biceps femoris.
    e.   peroneus.

7.   When contraction occurs:
    a.   the Z lines move closer together.
    b.   the A band remains constant.
    c.   the I band begins to disappear.
    d.   the H band gets smaller.
    e.   All of the above.

8.   The muscle group most lateral to the vertebral column is the:
    a.   spinalis.
    b.   intertransversarius.
    c.   iliocostalis.
    d.   longissimus.
    e.   multifidus.

9.   The most <u>superficial</u> layer of the abdominal muscles is the:
    a.   internal oblique.
    b.   transverses abdominis.
    c.   thoracis internus.
    d.   psoas major.
    e.   external oblique.

10. Actions of the *biceps brachii muscle* include:
    a.   pronation of the forearm.
    b.   flexion of the shoulder.
    c.   flexion of the elbow.
    d.   B and C
    e.   All of the above.

11. The powerful <u>flexors</u> of the <u>thigh</u> are the:
    a.   obturators.
    b.   iliopsoas.
    c.   gemelli.
    d.   pectineus.
    e.   adductors.

12. The connective tissue surrounding a muscle <u>fascicle</u> is called:
    a.   epimysium.
    b.   a tendon.
    c.   perimysium.

  d. endomysium.

  e. an aponeurosis.

13. The primary flexors of the wrist include which of the following muscles?

  a. supinator and brachioradialis

  b. flexor carpi ulnaris, flexor carpi radialis, and palmaris longus

  c. brachialis and palmaris longus

  d. A and C

  e. None of the above.

14. All the muscle fibers controlled by a single motor neuron constitute a:

  a. skeletal muscle.

  b. motor unit.

  c. ventral root.

  d. synaptic knob.

  e. sarcoplasmic reticulum.

15. Muscles in all of the following groups originate on the lower limb *except*:

  a. muscles that move the foot and toes.

  b. muscles that move the leg.

  c. muscles that move the thigh.

  d. B and C

  e. None of the above. (There are no exceptions.)

16. The function of the *tensor fascia lata* is to:

  a. steady the trunk on the thigh by making the iliotibial tract taut.

  b. stabilize the anterior aspect of the gluteus medius.

  c. assist in extension of the thigh.

  d. assist in medial rotation of the thigh and flexion of the knee.

  e. assist in flexion of the hip.

17. Which of the following muscles *cannot* compress the abdomen?

  a. internal oblique

  b. external oblique

  c. iliocostalis

  d. pectineus

  e. C and D

18. The action(s) of the *rhomboideus major muscle* include:

  a. abduct the scapula.

  b. adduct and rotate the scapula downward.

  c. depress and protract the shoulder.

  d. elevate the scapula.

  e. C and D

19. The sarcoplasmic reticulum stores:
    a. ATP.
    b. potassium ions.
    c. calcium ions.
    d. magnesium ion.
    e. A and C

20. The muscle that can both flex and adduct the arm is the:
    a. biceps brachii.
    b. subscapularis.
    c. pectoralis major.
    d. teres major.
    e. brachialis.

21. Which muscle(s), found on the dorsum of the foot, is/are found in the third layer?
    a. dorsal interossei
    b. abductor hallucis
    c. quadratus plantae
    d. flexor digitorum brevis
    e. adductor hallucis

22. Which muscle(s) acts to plantar flex the foot?
    a. gastrocnemius
    b. soleus
    c. plantaris
    d. A and B
    e. A, B, and C

23. Which hypothenar muscle(s) originates on the pisiform bone?
    a. flexor digiti minimi brevis
    b. abductor digiti minimi
    c. lumbricals
    d. opponens digiti minimi
    e. abductor pollicis brevis

24. The coronoid process of the ulna and the elbow joint capsule comprise the site of insertion for which muscle?
    a. brachialis
    b. biceps brachii
    c. brachioradialis
    d. triceps brachii
    e. coracobrachialis

25. Which of the following is *not* a function of the *teres minor* muscle?
    a. rotates the humerus laterally
    b. helps hold the humerus in the glenoid cavity
    c. is a synergist of the pectoralis major muscle
    d. adduction at the shoulder
    e. stabilizes the shoulder joint

26. Which muscles act to elevate the hyoid bone and steady it during swallowing and speech?
    a. sternohyoid
    b. thyrohyoid
    c. sternothyroid
    d. omohyoid
    e. digastric

27. Which of the following is *not* a characteristic of the *medial pterygoid* muscle?
    a. protracts mandible
    b. closes jaw
    c. promotes side-to-side (grinding) movements
    d. functions during mastication
    e. None of the above. They are all characteristics of the pterygoid muscle.

28. Which of the following is/are <u>true</u> regarding white (type IIx) 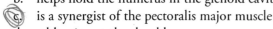 fibers?
    a. They depend on anaerobic pathways to make ATP.
    b. They contract rapidly.
    c. They fatigue quickly.
    d. They contain little myoglobin; thus, they are pale.
    e. All of the above are true.

Chapter 12 - THE NERVOUS SYSTEM I: NERVOUS TISSUE

INTRODUCTION

- The nervous system is one of the smallest, yet the most complex organ systems in the human body. It includes all of the neural tissue in the body.

- The nervous system, along with the endocrine system, controls and adjusts the activities of most of the other organ systems in the body. Its chief functions are to MONITOR, INTEGRATE, and RESPOND to information (*stimulus*; pl. = *stimuli*) in the environment.

A.   ANATOMICAL SUBDIVISIONS of the NERVOUS SYSTEM

- **Central Nervous System (CNS):** consists of the BRAIN and SPINAL CORD

- **Peripheral Nervous System (PNS):** consists of NERVES and GANGLIA, which are all the nervous tissue structures external to the CNS

- The nervous system RECEIVES SENSORY INPUTS and DICTATES MOTOR OUTPUTS. The two **FUNCTIONAL DIVISIONS** are:

  - **AFFERENT (SENSORY) DIVISION**: sensory impulses are carried from sensory receptors through the PNS TOWARD the CNS.

  - **EFFERENT (MOTOR) DIVISION**: motor impulses are carried AWAY FROM the CNS, through the PNS, to the EFFECTORS, which are muscles and glands. It is further subdivided into:

    - **SOMATIC NERVOUS SYSTEM (SNS)** – provides voluntary control over skeletal muscle contraction

    - **AUTONOMIC NERVOUS SYSTEM (ANS)** – provides automatic control, involving regulation of smooth muscles, cardiac muscle, and glandular activity

- The TYPES OF SENSORY INPUTS and MOTOR OUTPUTS are categorized as:

  - SOMATIC – refers to the OUTER body

- VISCERAL – refers to mainly the INNER body

- GENERAL - widespread

- SPECIAL – localized

- BRANCHIAL innervation – refers to the motor innervation of the pharyngeal (branchial) muscle

- PROPRIOCEPTION – refers to a series of senses that monitor the degree of stretch in muscles, tendons, and joint capsules. Therefore, this refers to SENSING the POSITIONS and MOVEMENTS of our body parts

B. | **NERVOUS (NEURAL) TISSUE ORGANIZATION** | : comprised of **neurons** and supporting cells called **neuroglia** or **glial cells**

- | **THE NEURON** |

  - Neurons are long-lived, non-dividing cells. Each has a **CELL BODY** (or **SOMA**) and cell processes, called **AXONS** and **DENDRITES**.

  - The neuron cell SOMA contains a nucleus surrounded by cytoplasm (called **nucleoplasm or perikaryon**), which contains supportive **neurofibrils, neurotubules, neurofilaments** and **chromatophilic (Nissl) bodies.**

    - **NISSL BODIES:** concentrations of RER and free ribosomes.

    - **EXCEPT for those found IN GANGLIA of the PNS, ALL NEURON CELL BODIES ARE IN THE CNS.**

  - **AXON FEATURES:**

    - **AXON HILLOCK:** specialized region of an axon, which connects the initial segment of the axon to the cell body

    - **AXOPLASM:** cytoplasm of axon, which contains numerous organelles

    - **COLLATERALS:** side branches from an axon

    - **TERMINAL ARBORIZATIONS:** a series of fine, terminal extensions, which branch from the tip of the axon and end at synaptic terminals

    - **TERMINAL BOUTON:** the area where one neuron synapses on another

    - **AXOLEMMA:** plasmalemma of an axon

- **IN THE PNS:**

  - **GANGLIA:** clusters of PNS neuron cell bodies

  - **NERVES:** bundles of axons in the PNS

- Most neurons have a number of branched **DENDRITES**, which are RECEPTIVE SITES that conduct signals from other neurons TOWARD the neuron cell body.

- Most neurons have one **AXON**, which generates and conducts nerve impulses AWAY FROM the neuron cell body.

- **SYNAPSE:** a functional junction between neurons (or between a neuron and another cell) at neuroeffector junctions

  - Functions as a **site of intercellular communication**

  - Occurs on dendrites (AXODENDRITIC), the cell body (AXOSOMATIC), or along axons (AXOAXONIC)

  - **VESICULAR SYNAPSE (chemical synapse):** involves neurotransmitters

  - **NONVESICULAR SYNAPSE (electrical synapse):** involves direct contact between cells

- **Anatomically, neurons are classified by the number of processes issuing from their cell body as:**

  - MULTIPOLAR – includes several dendrites and one axon

  - BIPOLAR – includes one dendrite and one axon

  - UNIPOLAR - contains one process

  - PSEUDOUNIPOLAR – the dendrite and axon are continuous at one side of the cell body

  - ANAXONIC – contains no distinguishable axon

- **Functionally, neurons are classified according to the direction in which they conduct impulses:**

  - **SENSORY (afferent) NEURONS:** conduct impulses TOWARD the CNS o **MOTOR (efferent) NEURONS:** conduct impulses AWAY FROM the CNS

o **INTERNEURONS (association neurons):** lie in the CNS between sensory and motor neurons

---

• | **NEUROGLIA and OTHER SUPPORTING CELLS** |

• There are non-neural supporting cells in neural tissue, which support, protect, nourish and insulate neurons.

• **NEUROGLIA or GLIAL CELLS:** supporting cells of the nervous system

NEUROGLIA of the CNS include:

o **ASTROCYTES** – the largest and most numerous of the glial cells; functions in controlling the interstitial environment, repairing damaged neural tissue, creating a 3-D framework for the CNS, guiding neuron development, and maintaining the blood-brain barrier (BBB), which isolates the CNS from the environment (*discussed further in CNS section*).

o **MICROGLIA** – phagocytic cells of the CNS (engulf cellular debris, waste products, and pathogens)

o **EPENDYMAL CELLS** – cuboidal to columnar epithelial cells that line the central canal and ventricles of the brain

o **OLIGODENDROCYTES** - glial cells responsible for maintaining cellular organization in the gray matter and producing myelin to completely sheath areas of white matter (myelinated axons)

NEUROGLIA or Supporting cells of the PNS:

o **SCHWANN CELLS** (*neurolemmocytes*): myelin-forming cells that cover all peripheral axons, whether myelinated or unmyelinated

o **SATELLITE CELLS:** enclose neuron cell bodies in the peripheral ganglia; regulate the exchange of nutrients and waste products between the neuron cell body and the extracellular fluid

- Thick axons are myelinated. **MYELIN speeds impulse conduction** along these axons.

- **MYELIN SHEATH** = a coat of supporting-cell membranes wrapped in layers around the axon. The sheath has gaps called **NODES of RANVIER** (or *neurofibral nodes*). Unmyelinated axons are surrounded by supporting cells, but they are not wrapped by layers of myelin.

- **PERIPHERAL NERVES**

  - A *peripheral nerve*, or simply *nerve*, is a BUNDLE of AXONS, wrapped in connective tissue, in the PNS.

    o Each AXON is enclosed by an **ENDONEURIUM**.

    o Each FASCICLE of axons is wrapped by a **PERINEURIUM**.

    o The WHOLE NERVE is surrounded by the **EPINEURIUM**.

  - Nerves are organs because they contain more than one kind of tissue.

- **REFLEX ARCS**

  - REFLEXES are rapid, automatic responses to stimuli, which can be either somatic or visceral.

  - There are 5 minimum numbers of elements in a reflex arc:

    o **RECEPTOR**
    o **SENSORY NEURON**
    o **INTEGRATION CENTER**
    o **MOTOR NEURON**
    o **EFFECTOR**

- ### SIMPLIFIED DESIGN of the NERVOUS SYSTEM

  - SIMPLE 3-NEURON REFLEX ARCS form the basis of the structural plan of the entire nervous system.

    - **SENSORY NEURONS enter the spinal cord dorsally.**

    - **MOTOR AXONS exit it ventrally.**

    - **INTERNEURONS are confined to the CNS.**

    - **The nerves in the PNS consist of the peripheral axons of the sensory and motor neurons.**

    - The cell bodies of motor neurons and interneurons make up the internal gray matter of the CNS, whereas the cell bodies of sensory neurons lie external to the CNS in sensory ganglia of the PNS.

  - Throughout most of the CNS, the inner gray matter is surrounded by outer white matter. The extreme center of the spinal cord and brain is a fluid-filled hollow central cavity.

Chapter 13 – THE NERVOUS SYSTEM II:
THE CENTRAL NERVOUS SYSTEM (CNS)

A. The BRAIN

- Provides for voluntary and involuntary movements

- Functions in interpretation and integration of sensation

- Provide consciousness and cognitive functions

- Also involved in innervation of the head through the cranial nerves

- **BASIC ORGANIZATION of the BRAIN** (4 basic parts)

  o Cerebral Hemispheres (Cerebrum)
  o Diencephalon
  o Brain Stem (Midbrain, Pons, Medulla)
  o Cerebellum

- INTERNAL STRUCTURE

  - **GRAY MATTER** = mostly CELL BODIES of neurons; some short <u>unmyelinated</u> axons and dendrites; some neuroglia

  - **WHITE MATTER** = mostly myelinated AXONS (or fibers) of neurons

  - BRAIN STEM: white matter external to central gray matter

  - CEREBRUM and CEREBELLUM: additional external cortex of gray matter

- VENTRICLES of the BRAIN

  - Ventricles are expansions of the brain's central cavity, filled with cerebrospinal fluid (CSF) and lined by ependymal cells; CSF functions in floating and cushioning the brain and the spinal cord.

  - **LATERAL VENTRICLES (2):** span both the cerebral hemispheres, separated by the SEPTUM PELLUCIDUM

- **THIRD (3ʳᵈ) VENTRICLE**: enclosed by the diencephalon

- **CEREBRAL AQUEDUCT**: located in the midbrain; connects the 3ʳᵈ and 4ᵗʰ ventricles

- **FOURTH (4ᵗʰ) VENTRICLE**: located in the hindbrain (dorsal to pons and superior ½ of medulla); continues inferiorly as the central cavity in the spinal cord

- **CHOROID PLEXUS**: the vascular complex in the roofs of the 3ʳᵈ and 4ᵗʰ ventricles; responsible for CSF production

The CEREBRAL HEMISPHERES (CEREBRUM)

- **FEATURES** : Grooves on and around the hemispheres

  - **FISSURES:** deepest grooves, which separate major portions of the brain

    - TRANSVERSE CEREBRAL FISSURE: separates cerebrum from cerebellum inferiorly

    - LONGITUDINAL FISSURE: separates the right (**R**) and left (**L**) cerebral hemispheres

  - **SULCI** (singular: **SULCUS** = "*furrow*"): the many grooves on the surface

  - **GYRI** (singular: **GYRUS** = "*twister*"): twisted ridges of brain tissue (*"lumpy bumps"*)

- EACH CEREBRAL HEMISPHERE has **FIVE MAJOR LOBES** , SEPARATED by SULCI:

  - **FRONTAL LOBE**: separated from parietal lobe by the **CENTRAL SULCUS**

  - **PARIETAL LOBE**

  - **OCCIPITAL LOBE**: lies farthest posteriorly; separated from parietal lobe by the **PARIETO-OCCIPITAL SULCUS**

- **TEMPORAL LOBE**: lateral side of hemisphere; separated from the parietal and frontal lobes by the **LATERAL SULCUS**; inferiorly separated from the occipital lobe by the **CALCARINE SULCUS**

- **INSULAR LOBE (or INSULA)**: buried deep within the lateral sulcus and forms part of its floor; covered by parts of the temporal, parietal and frontal lobes

- **INTERNAL STRUCTURE of CEREBRUM** : consists of the three largest regions within the cerebrum

  - CEREBRAL CORTEX of GRAY MATTER (*superficial layer*)
  - CEREBRAL WHITE MATTER (*internal*)
  - BASAL NUCLEI (or Basal Ganglia): *deep* in the white matter

- **CEREBRAL CORTEX**

  - Site of conscious sensory perception, voluntary initiation of movements, higher thought functions

  - Contains billions of neurons arranged into SIX LAYERS

  - Brodmann Areas: 52 structurally different areas identified by Korbinian Brodmann

There are three FUNCTIONAL AREAS of the CEREBRAL CORTEX: *MOTOR / SENSORY / ASSOCIATION AREAS*

  - **MOTOR AREAS**

    o **PRIMARY (1°) MOTOR CORTEX** (1° Motor Area or Somatic Motor Area)

      - Located along the PRECENTRAL GYRUS (Area 4)

      - The pyramidal axons in this area signal motor neurons to bring about PRECISE or SKILLED VOLUNTARY MOVEMENTS of the body, especially the forearms, fingers and facial muscles.

      - CONTRALATERAL PROJECTIONS: R and L 1° motor cortices (*plural of cortex*) control muscles on the L and R sides of the body, respectively

      - **MOTOR HOMUNCULUS** : body map on the motor cortex

- The <u>motor neurons</u> for the human body are represented spatially in the 1° motor cortex of each hemisphere. The pyramidal cells or large motor neurons that control hand movement are in one place; those that control foot movement are in another, etc.

- THE BODY IS REPRESENTED UPSIDE DOWN.

- The FACE and HAND REPRESENTATIONS are DISPROPORTIONATELY LARGE in order to provide more pyramidal cells to control for the delicate and skilled movements.

- SOMATOTOPY: general principle of "body mapping"

o **PREMOTOR CORTEX**: anterior to the precentral gyrus (Area 6)

- CONTROLS MORE COMPLEX MOVEMENTS than does the 1° motor cortex

o **FRONTAL EYE FIELD**: anterior to the premotor cortex (Area 8)

- CONTROLS VOLUNTARY MOVEMENTS of the EYES

o **BROCA'S AREA**: anterior to the inferior part of the premotor cortex; **in the L cerebral hemisphere only** (overlaps Areas 44 and 45)

- MANAGES SPEECH PRODUCTION; controls movements necessary for speaking (corresponding region in the R hemisphere controls the emotional overtones given to speech)

- **SENSORY AREAS** : located in the parietal, temporal and occipital lobes

Cortical areas involved with CONSCIOUS AWARENESS of SENSATION

THERE IS A **DISTINCT CORTICAL AREA FOR EACH OF THE MAJOR SENSES** (some ASSOCIATION AREAS INCLUDED as well)

o **PRIMARY (1°) SOMATOSENSORY CORTEX**: located along the POSTCENTRAL GYRUS (Areas 3, 1, 2)

- Involved with CONSCIOUS AWARENESS of the GENERAL SOMATIC SENSES

- SPATIAL DISCRIMINATION

- CONTRALATERAL PROJECTIONS

- SOMATOTOPY exhibited

- **SENSORY HOMUNCULUS** :

    - As described for motor homunculus, but <u>sensory neurons</u> are involved

    - The LIPS and HANDS are the most sensitive body parts – therefore, the REPRESENTATIONS of both on the homunculus are DISPROPORTIONATELY LARGE.

○ **SOMATOSENSORY ASSOCIATION CORTEX/AREA:** lies posterior to 1° somatosensory cortex (Areas 5 and 7)

- INTEGRATES DIFFERENT SENSORY INPUTS (touch, pressure, etc.) into a comprehensive understanding of WHAT IS BEING FELT

○ **VISUAL AREAS** : Overall, ~30 cortical areas are involved in visual processing, involving the occipital, temporal and parietal lobes

- **1° VISUAL (STRIATE) CORTEX:** largest cortical sensory area; on the posterior and medial part of occipital lobe; <u>most of it is located within the deep calcarine sulcus on medial aspect of occipital lobe</u> (Area 17)

    - RECEIVES VISUAL INFO that ORIGINATES on the RETINA of the EYE

- **VISUAL ASSOCIATION AREA:** surrounds the 1° visual area and covers much of the occipital lobe (Areas 18 and 19)

    - COMMUNICATES with 1° VISUAL AREA; CONTINUES PROCESSING of VISUAL INFO by ANALYZING COLOR, FORM and MOVEMENT

○ **AUDITORY AREAS**

- **1° AUDITORY CORTEX:** superior edge of temporal lobe, primarily inside the lateral sulcus (Areas 41 and 42)

    - CONSCIOUS AWARENESS of SOUND (loudness, rhythm and pitch)

- **AUDITORY ASSOCIATION AREA:** lies just posterior to 1° auditory cortex (posterior part of Area 22)

- - PERMITS EVALUATION of a SOUND (e.g. screech, thunder, music, etc.)

  - This area is usually located in the LEFT HEMISPHERE, in the CENTER of WERNICKE'S AREA (functional brain region involved in recognizing and understanding spoken words)

- **GUSTATORY CORTEX** : lies on the roof of the lateral sulcus (Area 43)

  - CONSCIOUS AWARENESS of TASTE stimuli

- **VESTIBULAR (Equilibrium) CORTEX** : posterior part of insula

  - CONSCIOUS AWARENESS of the SENSE of BALANCE

- **OLFACTORY CORTEX** : medial aspect of cerebrum in the piriform lobe (functional area for olfaction), which is dominated by the hook-like *uncus* (a medial, or inner, protrusion near the ventral surface of the temporal lobe)

  - CONSCIOUS AWARENESS of SMELL (*olfaction*)

- **ASSOCIATION AREAS** : "higher-order processing areas"

  Some of these areas make associations between (or tie together) the different kinds of sensory information received

  - **PREFRONTAL CORTEX**: large region of the frontal lobe, anterior to motor areas

    - The MOST COMPLICATED CORTICAL REGION

    - PERFORMS MANY COGNITIVE FUNCTIONS

  - **GENERAL INTERPRETATION AREA**: posterolateral cerebral cortex at the interface of the visual, auditory and somatosensory association areas

    - INTEGRATES ALL these TYPES of SENSORY INFORMATION

  - **LANGUAGE AREA** : a complex of functional areas that surround the lateral sulcus in the LEFT HEMISPHERE

- INVOLVED IN VARIOUS FUNCTIONS RELATED TO LANGUAGE

- FIVE SECTIONS of this area identified:
  1. **BROCA'S AREA:** speech production (**MOTOR**)

  2. **WERNICKE'S AREA:** speech comprehension (**SENSORY**)

  3. **LATERAL PREFRONTAL CORTEX:** deep conceptual <u>analysis</u> of spoken words

  4. **MOST of the LATERAL and INFERIOR TEMPORAL LOBE:** <u>coordination</u> of auditory and visual aspects of language

  5. **PARTS of the INSULA:** <u>initiation</u> of word articulation and <u>recognition</u> of rhymes and sound sequences

---

Corresponding areas on the RIGHT HEMISPHERE act in the CREATIVE INTERPRETATION of WORDS and in CONTROLLING EMOTIONAL OVERTONES of SPEECH. This CORRESPONDING AREA is NOT INVOLVED in the MECHANICS of SPEECH.

---

o **INSULA** : its function is not well understood

  - SOME PARTS FUNCTION in LANGUAGE and in the SENSE of BALANCE

  - Other parts have VISCERAL FUNCTIONS, including conscious perception of visceral sensations and behavioral influences on cardiovascular activity

- **CEREBRAL WHITE MATTER**

The white matter of the cerebrum is comprised of many **axons (or fibers)** through which the different areas of the cerebral cortex extensively communicate. Most of these fibers are **myelinated and bundled into large tracts.**

<u>The fibers are classified, according to where they run:</u>
- **COMMISSURES:** composed of commissural fibers that RUN BETWEEN THE TWO HEMISPHERES

  o Interconnect corresponding gray areas of the R and L hemispheres

  o **CORPUS CALLOSUM:** the largest commissure; superior to lateral ventricles, deep within the longitudinal fissure

- **ASSOCIATION FIBERS**: CONNECT different cortical areas WITHIN THE SAME HEMISPHERE; run horizontally

- **PROJECTION FIBERS**: fibers that RUN VERTICALLY TO and FROM the BRAIN STEM and SPINAL CORD

  o They either descend from the cerebral cortex to more caudal parts of the CNS or ascend to the cortex from lower regions.

  o Through these fibers, sensory information reaches the cortex and motor instructions are then relayed to effectors.

  o **INTERNAL CAPSULE**: projection fibers that form a compact bundle b/w the thalamus and some of the basal nuclei

  o **CORONA RADIATA**: projection fibers to and from cerebral cortex, which fan out

- **BASAL NUCLEI (or BASAL GANGLIA)**

  - Paired masses of gray matter embedded deep within the cerebral white matter

  - Coordinates with the cerebral cortex to CONTROL COMPLEX MOVEMENTS

  - NUCLEI (NUCLEAR GROUPS):

    o **CAUDATE NUCLEUS**: arches superiorly over the thalamus; lies medial to internal capsule; **functions** in subconscious adjustment and modification of voluntary motor commands

    o **AMYGDALOID BODY or NUCLEUS (*amygdala*)**: on the tip of the tail of the caudate nucleus; but, it functionally belongs to the limbic system

    o <u>Three masses of gray matter (CLAUSTRUM, PUTAMEN, and GLOBUS PALLIDUS) lie between the bulging surface of the insula and the lateral wall of the diencephalon.</u>

      - **CLAUSTRUM**: appears to be involved in visual information processing at the subconscious level, by focusing attention on specific patterns or relevant features

      - **LENTIFORM NUCLEI**: consists of the medial **GLOBUS PALLIDUS** and the lateral **PUTAMEN; functions as above** for caudate nucleus

      - **CORPUS STRIATUM**: caudate + lentiform nuclei

The DIENCEPHALON

- Forms the central core of the forebrain; surrounded by the cerebral hemispheres

Consists of 3 paired structures, enclosing the 3rd ventricle:

- **THALAMUS** (plural = *thalami*): egg-shaped or football-shaped

    - Both the R and L thalami form the walls of the diencephalon and the superolateral walls of the 3rd ventricle.

        - **INTERTHALAMIC ADHESION (or *massa intermedia*)** – a medial projection of gray matter, from the thalamus on either side, that extends into the 3rd ventricle; the two intermediate masses **fuse in the midline**, interconnecting the R and L thalami (occurs in ~70% of the population)

    - The thalamic nuclei **provide the switching and relay centers for both sensory and motor pathways**. (*Exception*: <u>sensory information</u> from the olfactory nerve and from the spinocerebellar tracts – NOT processed in the thalamic nuclei before the information is relayed to the cerebrum or brain stem.)

    - Contains about a dozen major nuclei, which send axons to particular portions of the cerebral cortex; concerned primarily with the relay of sensory information to the basal nuclei and cerebral cortex; The FIVE MAJOR GROUPS of NUCLEI are:

        - <u>ANTERIOR NUCLEI</u> – part of the limbic system; role in emotions, memory, and learning

        - <u>MEDIAL NUCLEI</u> – provide a conscious awareness of emotional states

        - <u>VENTRAL NUCLEI</u> – relay information to and from the basal nuclei and cerebral cortex

            - VENTRAL ANTERIOR – relay information regarding somatic motor commands form the basal nuclei and cerebellum to the 1° motor cortex and premotor cortex

            - VENTRAL LATERAL – as above

            - VENTRAL POSTERIOR – relay sensory information concerning tough, pressure, pain, temperature, and proprioception from the spinal cord and brain stem to the 1° sensory cortex of the parietal lobe

- o POSTERIOR NUCLEI

  - PULVINAR – integrates sensory information for projection to the association areas of the cerebral cortex

  - LATERAL GENICULATE NUCLEI (LGN) – receives visual information from the eyes, via the optic tract

  - MEDIAL GENICULATE NUCLEI (MGN) – relay auditory information to the auditory cortex from the specialized receptors of the inner ear

  - o LATERAL NUCLEI – relay stations in feedback loops that adjust activity in the cingulate gyrus and parietal lobe

  - SOME NUCLEI act as **RELAY STATIONS (processing centers) for SENSORY INFORMATION**, which ascend to the 1° sensory areas of the cerebral cortex.

    - o e.g. Ventral posterolateral nuclei receive general somatic sensory information; LGN and MGN receive visual and auditory information, respectively.

  - Every part of the brain that communicates with the cortex must relay its signals through a nucleus of the thalamus.

- **HYPOTHALAMUS** : inferior portion of the diencephalon

  - **The most important VISCERAL CONTROL CENTER**

    - o Contains centers involved with emotions and visceral processes that affect the cerebrum and other components of the brain stem

    - o Forms the link between the nervous and endocrine systems

  - **Controls a variety of autonomic functions**

    - o Regulates sleep cycles, hunger, thirst, body temperature, secretion of the pituitary gland and the autonomic nervous system

    - o Regulates some emotions and behaviors

  - Lies between the **OPTIC CHIASMA** or **OPTIC CHIASM** (where the optic tracts from the eyes cross) and the posterior border of the **MAMILLARY BODIES** (*see Limbic System section*)

- **INFUNDIBULUM:** located posterior to the optic chiasma, it connects the hypothalamus to the pituitary gland

- The PITUITARY GLAND projects inferiorly from the hypothalamus and secretes many hormones.

- **TUBERAL AREA:** the floor of the hypothalamus (between the infundibulum and the mammillary bodies), which contains nuclei involved with the control of pituitary gland function

- The hypothalamus forms the inferolateral walls of the 3rd ventricle.

- The hypothalamus contains roughly a dozen nuclei of gray matter.

  - **SUPRAOPTIC NUCLEUS** (SON) – produces antidiuretic hormone (ADH), which restricts water loss at the kidneys

  - **PARAVENTRICULAR NUCLEUS** (PVN) – produces oxytocin, which stimulates smooth muscle contractions in the uterus and prostate gland, and myoepithelial cell contractions in the mammary glands

  - **PREOPTIC AREA** – controls physiological responses to changes in body temperature

  - **SUPRACHIASMATIC NUCLEUS** (SCN) - the "body's biological clock," which regulates the timing of many daily (circadian) rhythms (e.g. light-dark or day-night cycles); its output adjusts the activities of other hypothalamic nuclei, the pineal gland, and the reticular formation

- **EPITHALAMUS** : most dorsal part of diencephalon

  - Forms part of the roof of the 3rd ventricle

  - **Consists of one tiny group of nuclei and the PINEAL GLAND (or PINEAL BODY),** which secretes the hormone, MELATONIN (involved in the regulation of day-night cycles -- signals the body to prepare for the nighttime stage of sleep-wake cycle)

The BRAIN STEM

- **GENERAL FUNCTIONS:**

  - Produces the rigidly programmed, **automatic behaviors** necessary for our survival

- Acts as a **passageway** for all the fiber tracts running between the cerebrum and the spinal cord

- Heavily involved with **innervation of the face and head**: 10 of the 12 cranial nerves attach to it

- **INTERNAL STRUCTURE:**
  - INNER region of GRAY MATTER
  - EXTERNAL WHITE MATTER
  - NUCLEI of GRAY MATTER located <u>within white matter</u>

- **3 BRAIN STEM REGIONS** from ROSTRAL (towards the snout or head) to CAUDAL (towards the tail end): *Midbrain (Mesencephalon) – Pons – Medulla Oblongata (Medulla)*

  - **MIDBRAIN (or MESENCEPHALON)**

    o Lies between the diencephalon and the pons

    o Its central cavity, CEREBRAL AQUEDUCT, divides it into a TECTUM (roof) dorsally and a PAIR of CEREBRAL PEDUNCLES ventrally.

    o **CEREBRAL PEDUNCLES**: pyramidal fiber tracts that form vertical pillars, which appear to "hold up the forebrain"

      - Contain ascending fibers that synapse in the thalamic nuclei

      - Contain descending fibers of the corticospinal pathway that carry voluntary motor commands from 1° the motor cortex of each cerebral hemisphere

    o **CORPORA QUADRIGEMINA**: large group of nuclei (appears as four bumps on the dorsal surface) that make up the tectum, which act in the "startle response" – process visual and auditory information and generate reflexive responses to these stimuli

      - **SUPERIOR COLLICULI** (sing. = *colliculus*): paired nuclei that act in <u>VISUAL REFLEXES</u>

      - **INFERIOR COLLICULI** (sing. = *colliculus*): paired nuclei that act in <u>AUDITORY REFLEXES</u>

○ Contains the **major nuclei of the reticular formation, RF,** (*discussed later*); each side of the midbrain contains a pair of the following nuclei:

- **RED NUCLEUS** – nucleus with abundant blood vessels, giving it a rich red coloration -- hence, its name; integrates information from the cerebrum and cerebellum and issues involuntary motor commands concerned with the **maintenance of muscle tone and limb position**

- **SUBSTANTIA NIGRA** - lies lateral to the red nucleus; gray matter contains darkly pigmented cells, giving it a black coloration; plays an **important role in regulating the motor output of the basal nuclei**

- **The PONS**

○ Anteriorly, it forms a bulging region that is wedged b/w the midbrain (mesencephalon) and the medulla oblongata.

○ Dorsally separated from the cerebellum by the 4th ventricle

○ Forms a VENTRAL BRIDGE b/w the R and L HALVES of the CEREBELLUM

- On either side, the pons is attached to the cerebellum by three **CEREBELLAR PEDUNCLES** (SUPERIOR, MIDDLE, and INFERIOR CEREBELLAR PEDUNCLES – *see Cerebellum Section*)

○ **PONTINE NUCLEI:** relay nuclei by which the motor cortex communicates with the cerebellum; involved in COORDINATION of VOLUNTARY MOVEMENTS

○ **NUCLEI concerned with the INVOLUNTARY CONTROL of RESPIRATION:** on each side of the brain, the RF in this region contains two respiratory centers (APNEUSTIC CENTER and PNEUMOTAXIC CENTER), which modify the activity of the RESPIRATORY RHYTHMICITY CENTER in the medulla oblongata

- **MEDULLA OBLONGATA (or MEDULLA)**

○ Most caudal part of the brain stem; continuous with the spinal cord at level of the foramen magnum; therefore, it physically connects the brain with the spinal cord – many of its functions are directly related to this connection

○ **MEDULLARY PYRAMIDS** (formed by the pyramidal tracts) flank the ventral midline.

o **MEDULLARY OLIVES** lie lateral to each pyramid and contain the **INFERIOR OLIVARY NUCLEI** (relay station for sensory information going to the cerebellum).

o **Other nuclei in the medulla oblongata are grouped into the following** **FUNCTIONS** .

- **RELAY STATIONS** : ascending tracts that synapse in sensory or motor nuclei act as relay stations and processing centers; some examples are:

   - **NUCLEUS GRACILIS** – relays somatic sensory information to the thalamus, from the lower body

   - **NUCLEUS CUNEATUS** – as above, afferents from the upper body

   - OLIVARY NUCLEI – relay sensory information from the spinal cord, the cerebral cortex, diencephalon, and brain stem to the cerebellar cortex

- **NUCLEI of CRANIAL NERVES** : sensory and motor nuclei associated with five of the cranial nerves (VIII, IX, X, XI, and XII), which innervate muscles of the pharynx, neck, and back, and also visceral organs of the thoracic and peritoneal cavities

- **AUTONOMIC NUCLEI** : RF nuclei and centers that are responsible for the regulation of vital autonomic functions; major centers include the following:

   - **CARDIOVASCULAR CENTERS** – adjust heart rate, the strength of cardiac contractions, and the blood flow through peripheral tissues

   - **RESPIRATORY RHYTHMICITY CENTERS** – set the basic pace for respiratory movements; their activity is regulated by inputs from the *apneustic* and *pneumotaxic centers* of the pons

The CEREBELLUM

- **FUNCTIONS** :

   - SMOOTHS and COORDINATES BODY MOVEMENTS that are directed by other brain regions

   - HELPS MAINTAIN POSTURE and EQUILIBRIUM

- **GROSS STRUCTURE** :

  - 2 CEREBELLAR HEMISPHERES connected medially by the **VERMIS** (worm-like strx)

  - Its surface is folded into **FOLIA** (plate-like ridges – less prominent than the *gyri*, singular = *gyrus*, of the cerebral hemispheres), which are separated by **FISSURES** (grooves)

    o **PRIMARY FISSURE** – separates the two major lobes (anterior and posterior lobes) of each hemisphere

  - Each cerebellar hemisphere is divided into 2 major lobes + 1 slender lobe :
    o **ANTERIOR LOBE** – lies superior to posterior lobe; assists, along with the posterior lobe, in the planning, execution, and coordination of limb and trunk movements

    o **POSTERIOR LOBE** – lies inferior to anterior lobe and posterior to flocculonodular lobe

    o **FLOCCULONODULAR LOBE** – lies anterior and inferior to the cerebellar hemisphere; important role in the maintenance of balance and the control of eye movements

- **INTERNAL STRUCTURE of the CEREBELLUM** : 3 regions in X-section

  - OUTER CEREBLLAR CORTEX (neural cortex) of GRAY MATTER, which contains large, highly branched **PURKINJE CELLS**; **functions** in subconscious coordination and control of ongoing movements of body parts

  - INTERNAL WHITE MATTER
    o **ARBOR VITAE** (*"tree of life"*) – the white matter forms a branching array that resembles a tree in sectional view

  - DEEP **CEREBELLAR NUCLEI** of gray matter – a relatively small portion of the afferent (sensory) fibers synapse within these nuclei before projecting to the cerebellar cortex; **functions as above**, for cerebellar cortex

- **CEREBELLAR PEDUNCLES** : thick tracts of nerve fibers that connect the cerebellum to the brain stem

  - **SUPERIOR** CEREBELLAR PEDUNCLES: connects cerebellum to nuclei in the MIDBRAIN (mesencephalon), DIENCEPHALON, and CEREBRUM

- **MIDDLE** CEREBELLAR PEDUNCLES: connects cerebellum to PONS (including sensory and motor nuclei within), via a broad band of fibers that cross the ventral surface of the pons at right angles to the axis of the brain stem

- **INFERIOR** CEREBELLAR PEDUNCLES: connects cerebellum to nuclei in the MEDULLA OBLONGATA; carry ascending (sensory) and descending (motor) cerebellar tracts from the spinal cord

- Virtually all fibers that enter and leave the cerebellum are IPSILATERAL. (i.e. Damage/lesion in one side of the cerebellum affects the same side of the body.)

Memory aid: **CEREBRAL vs. CEREBELLAR PEDUNCLES**
CEREBRAL (cerebrum) PEDUNCLES are associated with the midbrain or mesencephalon – remember the analogy of how they appear to "hold up" the forebrain or cerebrum.
CEREBELLAR (cerebellum) PEDUNCLES are associated with both the pons and the cerebellum – remember that they link or bridge the pons to the cerebellum.

FUNCTIONAL BRAIN SYSTEMS

- The **LIMBIC SYSTEM** : "EMOTIONAL BRAIN"

  - Group of structures on **medial aspect of each cerebral hemisphere and the diencephalon**, forming a broad ring ("*limbus*" = headband), along the border between the cerebrum and diencephalon.

  - This functional system includes:
    - the **septal nuclei**
    - the **cingulate gyrus**
    - the **hippocampal formation** (*the hippocampus within the dentate gyrus and the parahippocampal gyrus*)
    - the **amygdaloid body**

  - The FORNIX (tract of white matter connecting the hippocampus with the hypothalamus) and other fiber tracts link the limbic system together.

    - **MAMMILLARY BODIES:** prominent nuclei in the floor of they hypothalamus, to which many of the fibers of the fornix end or connect; concerned with feeding reflexes and behaviors

  - **FUNCTIONS of the LIMBIC SYSTEM:**

    - Establishment of EMOTIONAL STATES and RELATED BEHAVIORAL DRIVES

- o LINK the CONSCIOUS, INTELLECTUAL FUNCTIONS of the CEREBRAL CORTEX with the UNCONSCIOUS and AUTONOMIC FUNCTIONS of OTHER PORTIONS of the brain

- o CONSOLIDATION (a function of MEMORY STORAGE) and RETRIEVAL of MEMORIES

- **The RETICULAR FORMATION (RF)**

  - A group of neurons that runs through the central core of the **medulla, pons and midbrain**.

  - CONTROLS AROUSAL of the BRAIN as a WHOLE due to the widespread connections of reticular neurons to other brain regions

  - The **RETICULAR ACTIVATING SYSTEM (RAS)** is a branch of the RF that maintains CONSCIOUSNESS and ALERTNESS.

  - The RAS also functions in sleep, and in arousal from sleep.

  - General anesthesia, alcohol, tranquilizers and sleep-inducing drugs depress the RAS.

The MENINGES

Comprised of three membranes of connective tissue that lie just external to the brain and spinal cord

- **FUNCTIONS:**

  - Cover and protect the CNS

  - Enclose and protect the blood vessels supplying the CNS

  - Contain cerebrospinal fluid (**CSF**)

- **LAYERS of the MENINGES:**
  DURA MATER – ARACHNOID MATER – PIA MATER (*External to Internal*)

  - **DURA MATER**: (*"tough mother"*) 2-layered sheet of fibrous connective tissue

    - o PERIOSTEAL LAYER: attaches to internal surface of skull bones

    - o MENINGEAL LAYER: forms true external covering of the brain

      - DURAL SINUSES: areas where both periosteal and meningeal layers separate to enclose these blood-filled sinuses, which ACT as VEINS

- FALX CEREBRI: ligamentous membrane that attaches anteriorly to crista galli of the ethmoid bone and STABILIZES the BRAIN WITHIN the CRANIAL CAVITY of the skull

- FALX CEREBELLI: ligamentous membrane that continues inferiorly from the posterior part of *falx cerebri* and runs along vermis of the cerebellum in the posterior cranial fossa

- **ARACHNOID MATER**:

  o SUBDURAL SPACE b/w the dura mater and the arachnoid mater; contains a film of CSF

  o SUBARACHNOID SPACE deep to arachnoid mater

    - Spanned by web-like threads that hold arachnoid mater to underlying pia mater

    - Filled with CSF and contains the largest blood vessels supplying the brain

  o **ARACHNOID VILLI act as valves** that allow CSF to pass from the subarachnoid space to dural sinuses

  o DURA MATER and ARACHNOID MATER SURROUND THE BRAIN LOOSELY.

- **PIA MATER**: (*"gentle mother"*) a layer of delicate connective tissue, richly vascularized with fine blood vessels

  o CLINGS TIGHTLY to the BRAIN SURFACE, following every convolution

CEREBROSPINAL FLUID (CSF)

- Watery broth located in and around the brain and the spinal cord

- Provides a liquid cushion or buoyancy to the CNS structures

- Helps nourish the brain

- Remove wastes produced by neurons

- Transmit neurotransmitters (chemical signals) b/w different parts of the CNS

- MOST of the CSF is made in the CHOROID PLEXUSES (vascular membranes in the roofs of the ventricles, mainly the third and fourth ventricles).

The BLOOD-BRAIN-BARRIER (BBB)

- TIGHT JUNCTIONS are special features of the endothelium of brain capillaries' walls, which help form a barrier b/w the brain and the rest of the body.

- The BBB is NOT an ABSOLUTE BARRIER, though, because all nutrients (including oxygen) and ions required by neurons can pass through the barrier.

- Lipid-soluble molecules such as alcohol, nicotine and anesthetics, can easily diffuse through the BBB and reach brain neurons.

B. The SPINAL CORD

- **FEATURES:**

  - The adult spinal cord extends ~45 cm (~18") from the foramen magnum to L_1 or L_2 vertebra.

  - It is protected by bone, meninges and cerebrospinal fluid (CSF).
    - SPINAL MENINGES – a group of specialized membranes that provides physical stability and shock absorption for the neural tissues of the spinal cord

  - **SPINAL DURAL SHEATH:** dura mater layer but does not attach to surrounding bone; corresponds to meningeal layer only

- **FUNCTIONS of the SPINAL CORD:**

  - Involved in SENSORY and MOTOR INNERVATION of the entire body INFERIOR to the HEAD

  - Provides a 2-WAY CONDUCTION PATHWAY for signals b/w the body and the brain

  - MAJOR CENTER for REFLEXES

- **GROSS STRUCTURE of the SPINAL CORD:**

  - There are **31 PAIRS of SPINAL NERVES** (the union of sensory and motor fibers distal to each dorsal root ganglion, DRG) that protrude through the *intervertebral foramina* (sing. = *foramen*), and attach to the spinal cord through the dorsal and ventral nerve roots. The 31 pairs are subdivided into:

    - **CERVICAL SPINAL NERVES (8 pairs)**
    - **THORACIC SPINAL NERVES (12 pairs)**
    - **LUMBAR SPINAL NERVES (5 pairs)**

- o **SACRAL SPINAL NERVES** (5 pairs)
- o **COCCYGEAL SPINAL NERVE** (1 pair)

> **Each of the 31 segments of the spinal cord is associated with a pair of DRG (contains sensory neuron cell bodies) and pairs of dorsal roots and ventral roots.** <u>EXCEPTION</u>: In many individuals, DORSAL ROOTS in the 1st cervical and 1st coccygeal nerves MAY NOT BE PRESENT.

- Each spinal nerve is ensheathed by a series of connective tissue layers, similar to those in muscle tissue:

  - o **EPINEURIUM** (outermost layer) – dense network of collagen fibers surrounding the entire nerve

  - o **PERINEURIUM** (middle layer) – partitions the nerve into fascicles and forms the NERVE-BLOOD BARRIER

  - o **ENDONEURIUM** (inner layer) – delicate connective tissue fibers that surround individual axons of the fascicles

- PERIPHERAL DISTRIBUTION of NERVES (*See PNS Section for further details*)

- **CERVICAL and LUMBAR ENGLARGEMENTS**: these enlarged areas of the anterior (ventral) horns of gray matter, (supplying the upper and lower limbs, respectively) arise from these enlargements

- **SPINAL MENINGES**

  DURA MATER – ARACHNOID MATER – PIA MATER (*External to Internal*)

  - o **DURA MATER** (*outermost layer*): tough, fibrous layer that covers the spinal cord, whose caudal end forms the COCCYGEAL LIGAMENT

    - **EPIDURAL SPACE**: external to dural sheath and separates the dura mater from the inner walls of the vertebral canal; filled with fat and veins; anesthetics often injected into this space

  - o **ARACHNOID MATER** (*middle layer*):

    - SUBDURAL SPACE – separates the dura mater from the arachnoid mater

    - SUBARACHNOID SPACE – internal to the arachnoid mater; contains the **ARACHNOID TRABECULAE**, a network of collagen and elastic fibers, and CSF

- CSF functions as shock absorber and diffusion medium for dissolved gases, nutrients, chemical messengers, and waste products

  o **PIA MATER** (*inner layer*)**:** firmly attached to the underlying neural tissue

  - | **DENTICULATE LIGAMENTS** | – comprised of supporting fibers (extending laterally from the spinal cord surface) that bind the spinal pia mater and arachnoid mater to the dura mater

    o Functions in preventing either lateral (side-to-side) or inferior movement of the spinal cord

- **CONUS MEDULLARIS**: inferior tapered end of spinal cord

- **FILUM TERMINALE** (*"end ligament"*): strand of fibrous connective tissue covered with pia mater

  o It extends through the vertebral canal to the S_2 vertebra.

  o It attaches to the coccyx inferiorly and anchors the spinal cord in place.

  o Ultimately, it becomes part of the *coccygeal ligament.*

- **CAUDA EQUINA** (*"horse's tail"*): consists of the lumbar and sacral nerve roots at the inferior end of vertebral canal, including the filum terminale

- **POSTERIOR MEDIAN SULCUS** (shallower groove) **and ANTERIOR MEDIAN FISSURE** (wide groove): two deep grooves that run the length of the spinal cord and PARTLY divide it into R and L halves

- | **INTERNAL STRUCTURE of the SPINAL CORD** |

  - **GRAY MATTER** of SPINAL CORD (*shaped like "H" in X-section*):

    o Surrounding the central canal is a central core of gray matter, which is a **mixture of motor neuron cell bodies, short unmyelinated axons and dendrites of sensory neurons, association neurons and neuroglia.**

o **GRAY COMMISSURE** (*crossbar of "H" in X-section*) posterior and anterior to the CENTRAL CANAL, this structure consists of axons of interneurons that cross from one side of the spinal cord to the other side

o **NUCLEI**: groups of neuron cell bodies in the spinal cord gray matter

o **POSTERIOR (DORSAL) HORNS**: consist of ALL interneurons

- Outside the spinal cord, the cell bodies are in the DORSAL ROOT GANGLIA and the axons are in the DORSAL ROOTS.

- Contains SOMATIC and VISCERAL SENSORY nuclei

o **ANTERIOR (VENTRAL) HORNS**: these areas are largest in the cervical and lumbar regions, which innervate the upper and lower limbs, respectively (cervical and lumbar enlargements).

- Provides SOMATIC MOTOR CONTROL

o **LATERAL HORNS**: present in the thoracic and superior lumbar segments of the spinal cord
- Consists of VISCERAL MOTOR NEURONS

- **DORSAL ROOTS (*sensory*) and VENTRAL ROOTS (*motor*)**: contain sensory and motor neuron fibers, respectively

o Dorsal and ventral roots <u>form the spinal nerves</u>. Therefore, spinal nerves are MIXED NERVES in that they contain both sensory and motor neuron fibers.

- **The WHITE MATTER on each side of the spinal cord is divided into three WHITE COLUMNS or FUNICULI** (sing. = *funiculus*), <u>named according to their positions</u>: (Each funiculus is comprised of bundles of tracts, called *fasciculi*; sing. = *fasciculus*.)

o **POSTERIOR FUNICULUS** (*dorsal white column*) is subdivided into:

- **FASCICULUS GRACILIS** (*medial*), which contains axon fibers supplying the <u>lower body</u>

- **FASCICULUS CUNEATUS** (*lateral*), which contains axon fibers supplying the <u>upper body</u>

o **ANTERIOR FUNICULUS** (*ventral white column*): continuous with lateral funiculus

- o **LATERAL FUNICULUS** (*lateral white column*)

  - **ASCENDING TRACTS:** groups of fibers that relay SENSORY information from the spinal cord to the brain

  - **DESCENDING TRACTS:** groups of fibers that relay MOTOR information from the brain to the spinal cord

Chapter 14 – THE NERVOUS SYSTEM III:
THE PERIPHERAL NERVOUS SYSTEM (PNS)

PART I : PNS INTRODUCTION and SPINAL NERVES

A. **INTRODUCTION**

- **PNS:** Nervous system structures OUTSIDE the BRAIN and SPINAL CORD; its nerves thread through almost every part of the body

- **NERVE:** a cord-like organ in the PNS consisting of many axons (nerve fibers) arranged in parallel bundles (fascicles), which are enclosed by successive wrappings of connective tissue

    - **EPINEURIUM:** external tough fibrous connective tissue sheath <u>surrounding a whole nerve</u>, which consists of several fascicles and blood vessels

        - **FASCICLES:** bundles of neuronal axons

    - **PERINEURIUM:** connective tissue <u>surrounding each fascicle</u>; forms the NERVE-BLOOD BARRIER

    - **ENDONEURIUM:** delicate connective tissue fibers that <u>surround the individual axons of fascicles</u>

- Most nerves are MIXED, carrying both sensory and motor axons.

B. **FUNCTIONAL COMPONENTS of the PNS**

- The CNS is connected to the PNS via **cranial nerves** and **spinal nerves**.

- The PNS is further subdivided into two divisions:

    - **SENSORY (AFFERENT) DIVISION**
        - SOMATIC Sensory
        - VISCERAL Sensory

    - **MOTOR (EFFERENT) DIVISION**
        - SOMATIC Motor
        - BRANCHIAL Motor
        - VISCERAL Motor, which comprise the AUTONOMIC NERVOUS SYSTEM (ANS), along with the Branchial Motor division

- ANS (*"involuntary nervous system"*) is further categorized into two functional divisions that serve most of the same organs but generally cause opposing or antagonistic effects:

    - PARASYMPATHETIC (*craniosacral, "Rest and Digest or Rest and Repose"*) DIVISION

    - SYMPATHETIC (*thoracolumbar; "Fight or Flight"*) DIVISION

C. **BASIC STRUCTURAL COMPONENTS of the PNS** *(Sensory Receptors / Motor Endings / Nerves and Ganglia)*

- **PERIPHERAL** **SENSORY RECEPTORS** : pick up stimuli (environmental changes) from inside and outside the body, then initiate impulses in sensory axons

    - **TWO MAIN CATEGORIES:**

        o DENDRITIC ENDINGS of sensory neurons

        o Complete RECEPTOR CELLS = specialized epithelial cells or small neurons that transfer special senses information

    - Receptors may be **classified by the location of their stimuli, the type of stimulus detected, and by their structure.** *(Refer to Chapter 16 for further details.)*

- **PERIPHERAL** **MOTOR ENDINGS** : axon terminals of motor neurons that innervate effectors (muscles and glands)

    - INNERVATION of SKELETAL MUSCLE

        o **NEUROMUSCULAR JUNCTIONS** (motor end plates): one junction associated with each muscle fiber

            - **Acetylcholine (ACh)** is the neurotransmitter (brain chemical) that diffuses across the synaptic cleft and binds to receptors on the sarcolemma; ACh then induces impulses, which signal the muscle cell to <u>contract</u>

        o **MOTOR UNIT**: a motor neuron innervating muscle fibers

- INNERVATION of VISCERAL MUSCLE and GLANDS

  o <u>Simpler arrangement</u>: Near the smooth muscle or gland cells that are innervated, a motor axon swells into a row of **VARICOSITIES** (knobs), which contain synaptic vesicles filled with neurotransmitters.

- **PERIPHERAL NERVES and GANGLIA**

  - Cranial nerves (CN) and spinal nerves

  - **GANGLION** (pl. = *ganglia*) : cluster of peripheral cell bodies
    DORSAL ROOT GANGLION (DRG): PNS ganglion containing the cell bodies of SENSORY neurons

  - **SPINAL NERVES**

    o **31 PAIRS** of spinal nerves arise from and span the length of the spinal cord; segmented into the following <u>paired divisions</u>:

      - **CERVICAL SPINAL NERVES (8 pairs)**
      - **THORACIC SPINAL NERVES (12 pairs)**
      - **LUMBAR SPINAL NERVES (5 pairs)**
      - **SACRAL SPINAL NERVES (5 pairs)**
      - **COCCYGEAL SPINAL NERVES (1 pair)**

    o The first branch of each spinal nerve in the thoracic (T_1) and upper lumbar (L_2) regions becomes the **WHITE RAMUS**, which contains myelinated **PRE**ganglionic axons that continue to an autonomic ganglion. <u>Two groups of **unmyelinated fibers** exit this ganglion</u>:

      - **GRAY RAMUS** (pl. = *rami*), which carries axons that innervate glands and smooth muscles in the body wall or limbs back to the spinal nerve

      - an **AUTONOMIC NERVE**, which carries fibers (axons) to internal organs

    o **RAMI COMMUNICANTES**: the white and gray rami, which collectively carry <u>visceral motor fibers</u> to and from a nearby autonomic ganglion associated with the sympathetic division of the ANS

    o **For $T_1 - L_2$, their spinal nerves have four branches:**

      - WHITE RAMUS

- GRAY RAMUS
- DORSAL RAMUS
- VENTRAL RAMUS

o **Each spinal nerve connects to the spinal cord via:**

- a **DORSAL ROOT**, which contains SENSORY FIBERS arising from cell bodies in DORSAL ROOT GANGLIA

- and a **VENTRAL ROOT**, which contains MOTOR FIBERS arising from cell bodies in the ANTERIOR (ventral) HORN of the SPINAL CORD

- Together these <u>sensory and motor fibers converge and exit the vertebral column</u> as **SPINAL NERVES**, which branch out as the DORSAL RAMUS and the VENTRAL RAMUS, respectively

  - **DORSAL RAMUS: supply dorsum of the neck and trunk, and specific segment of the skin**

  - **VENTRAL RAMUS: supply anterior (ventral) and lateral regions of neck and trunk, and all regions of the limbs**

  - Together, the DORSAL and VENTRAL RAMI supply SOMATIC REGIONS (skeletal musculature and skin) from the neck inferiorly

o Each pair of spinal nerves monitors a <u>specific region of the body surface</u>, an area known as:

- a **DERMATOME** – an area of the skin innervated by the cutaneous branches from a single spinal nerve

  - Clinically important because damage to either a spinal nerve or DRG will produce a characteristic loss of sensation in specific areas of the skin

- **NERVE PLEXUSES**

o Complex, interwoven networks of nerves <u>formed by the VENTRAL RAMI only</u>

o Occur as pairs in the cervical, brachial, lumbar and sacral regions

o PRIMARILY SERVE the LIMBS

o **CERVICAL PLEXUS** (Neck)

- Buried deep in the neck, under the sternocleidomastoid muscle

- Formed by the ventral rami of C_1 – C_4 and some fibers from C_5

- **CUTANEOUS NERVES:** supply only the skin of neck, back of head and most superior region of shoulder

- **PHRENIC NERVES:** supply the diaphragm

○ **BRACHIAL PLEXUS** (Upper Extremity or Limb)

- Lies partly in the neck and in the axilla (armpit)

- Formed by the ventral rami of C_5 - T_1

- Innervates the pectoral girdle and upper extremity

- Composed of <u>five consecutive groups of stems and branches, including the nerves they form</u>:

  - The **ROOTS** of the ventral rami of C_5 – T_1 converge to form the trunks

  - **TRUNKS** (SUPERIOR, MIDDLE, and INFERIOR Trunks) – each of which divides into an anterior division and a posterior division

  - **DIVISIONS** (ANTERIOR and POSTERIOR) – each division then converge to form cords

  - **CORDS**

    ○ LATERAL CORD – formed from the anterior divisions of the superior and middle trunks

    ○ MEDIAL CORD – formed by a continuation of the anterior division of the inferior trunk

    ○ POSTERIOR CORD – formed by the union of all three posterior divisions of the superior, middle, and inferior trunks

  - **NERVES** – arise from one or more trunks or cords whose names indicate their positions relative to the axillary artery, which supplies the upper limb

    ○ **MEDIAN NERVE** – formed by the lateral and medial cords

    ○ **MUSCULOCUTANEOUS NERVE** – formed exclusively by the lateral cord

○ **ULNAR NERVE** – formed by the medial cord

○ **AXILLARY NERVE** – formed by the posterior cord

○ **RADIAL NERVE** – formed by the posterior cord

Mnemonic device*:** To remember the progressive branching of the BRACHIAL PLEXUS, use the following phrase. ***"Really Tired? Drink Coffee Now!" corresponds to Roots – Trunks – Divisions – Cords - Nerves

---

○ **LUMBAR PLEXUS** (Lower Extremity or Limb)

- Lies within the *psoas major muscle* in the posterior abdominal wall

- Formed by the ventral rami of $T_{12} - L_4$

- Innervates the anterior thigh

- **FEMORAL NERVES:** innervate anterior thigh muscles, incl. quadriceps femoris

- **OBTURATOR NERVES:** innervate ADDUCTOR muscle group and some skin on superomedial thigh

○ **SACRAL PLEXUS** (Lower Extremity or Limb)

- Lies immediately caudal to the lumbar plexus

- Formed by the ventral rami of $L_4 - S_4$

- Innervates the buttock, lower limb, pelvis and perineum

- **SCIATIC NERVE:** thickest and longest nerve in the body; innervates all of the lower limb except anterior and medial thigh regions; actually composed of two nerves, wrapped in a common sheath

- **TIBIAL NERVE:** innervates almost all muscles in POSTERIOR LOWER LIMB

- **COMMON FIBULAR NERVE:** innervates ANTEROLATERAL aspect of LOWER LIMB

- **SUPERIOR / INFERIOR GLUTEAL NERVES:** innervate gluteal muscles

- **PUDENDAL NERVE:** innervate muscles and skin of perineum

PART II : THE CRANIAL NERVES

D. **INTRODUCTION to CRANIAL NERVES**

- 12 pairs of cranial nerves (CNs) attach to the brain and pass through various openings or *foramina* (singular = *foramen*) in the skull

- CN I – XII (rostral to caudal)

- CN I attaches to the forebrain; CN II – XII attach to the brain stem

- **CNs SERVE ONLY HEAD and NECK structures, *except CN X, which extends into the abdomen.***

- CNs contain SENSORY and MOTOR FIBERS that innervate the head

- CELL BODIES of SENSORY neurons lie either in receptor organs (e.g. nose, eye and ear) OR within cranial sensory ganglia, which lie along CN V, VII – X just external to the brain

- CELL BODIES of MOTOR neurons occur in CN nuclei in ventral gray matter of the brain stem

E. **FUNCTIONAL GROUPS** of CRANIAL NERVES (CN)

- **PURELY SENSORY NERVES (I, II, VIII):** consists of special somatic sensory fibers for smell, vision, hearing and equilibrium

- **PRIMARILY or EXCLUSIVELY MOTOR NERVES (III, IV, VI, XI, XII):** contain general somatic motor fibers to skeletal muscles of the eye and tongue

- **MIXED (motor and sensory) NERVES (V, VII, IX, and X):**

  - Consist of general <u>somatic sensory</u> fibers to the face

  - Consist of general <u>visceral sensory</u> fibers to the mouth, viscera and taste buds (special visceral sensory)

  - Consist of <u>branchial motor</u> fibers to all pharyngeal arch muscles (chewing muscles and muscles of facial expression)

  - AFFERENTS = SENSORY FIBERS

  - EFFERENTS = MOTOR FIBERS

F. | **CRANIAL NERVES MNEMONIC** | (*memory tool*): use the first letter of each word to remember the first letter of the name of each cranial nerve in the following list.

> It is advisable to LEARN the NAMES of EACH CN using the following *mnemonic devices*; learn, as well, the <u>main function of each nerve</u>, as summarized below, most of which are evident in the name of the nerve itself. Note that ROMAN NUMERALS are used, <u>not Arabic numbers</u>. Arabic numbers are used for the subdivisions of certain cranial nerves.

S - *CN I**   <u>O</u>LFACTORY NERVES (*sense of smell*)

S - *CN II**   <u>O</u>PTIC NERVES (*vision*)

M - CN III   <u>O</u>CULOMOTOR NERVES (*"eye mover" - innervate four extrinsic eye muscles*)

M - CN IV   <u>T</u>ROCHLEAR NERVES (*"pulley" - innervate the superior oblique muscles*)

B - CN V   <u>T</u>RIGEMINAL NERVES (*"three-fold" - sensory innervation of the face / motor fibers to chewing muscles*)

M - CN VI   <u>A</u>BDUCENS NERVES (*abducts the eyeball - innervate the lateral rectus muscles*)

B - CN VII   <u>F</u>ACIAL NERVES (*muscles of facial expression and other structures*)

S - *CN VIII**   **<u>A</u>UDITORY (VESTIBULOCOCHLEAR) NERVES (*hearing and equilibrium*)

B - CN IX   <u>G</u>LOSSOPHARYNGEAL NERVES (*tongue and pharynx*)

B - CN X   <u>V</u>AGUS NERVES (*pharynx, larynx, heart, lungs, abdominal viscera, etc.*)

M - CN XI   <u>A</u>CCESSORY NERVES (*accessory part of X – i.e. its internal branch joins CN X*)

M - CN XII   <u>H</u>YPOGLOSSAL NERVES (*tongue muscles*)

(*S* = *PURELY SENSORY NERVES / *M* = PRIMARILY MOTOR NERVES / *B* = BOTH SENSORY and MOTOR fibers in the nerves)

** Vestibulocochlear Nerve is currently the preferred name for CN VIII, but Auditory Nerve is an "old-school" term and is still acceptable.

> *Mnemonic devices* <u>for the names of CN I – XII in consecutive order</u>:
>
> 1) "<u>O</u>n <u>O</u>ld <u>O</u>lympus <u>T</u>owering <u>T</u>ops <u>A</u> <u>F</u>inn <u>A</u>nd <u>G</u>erman <u>V</u>iewed <u>A</u> <u>H</u>ouse."
>
> 2) "<u>O</u>h, <u>O</u>nce <u>O</u>ne <u>T</u>akes <u>T</u>he <u>A</u>natomy <u>F</u>inal, <u>A</u>ll <u>G</u>ood <u>V</u>acations <u>A</u>re <u>H</u>eavenly."
>
> 3) "<u>O</u>h, <u>O</u>h, <u>O</u>h, <u>T</u>o <u>T</u>ouch <u>A</u>nd <u>F</u>eel <u>V</u>ery <u>G</u>ood <u>V</u>elvet. <u>A</u><u>H</u>!"
>
> Version 3 refers to CN VIII as Vestibulocochlear Nerve, rather than Auditory Nerve.

> *Mnemonic device* <u>for the functional grouping of the CNs</u> *(sensory, motor, or both)*:
> "<u>S</u>ome <u>S</u>ay <u>M</u>arry <u>M</u>oney <u>B</u>ut <u>M</u>y <u>B</u>rother <u>S</u>ays <u>B</u>ig <u>B</u>rains <u>M</u>atter <u>M</u>ore."

G. THE CRANIAL NERVES

- ### OLFACTORY NERVES (CN I) (PURELY SENSORY)

  - Carry AFFERENT IMPULSES for sense of SMELL

  - **Arise from olfactory receptor cells** located in the olfactory epithelia of the nasal cavity

  - Project as **olfactory nerve filaments**, which pass through the cribriform plate of ethmoid bone

  - ANOSMIA: partial or total loss of smell caused by fractured ethmoid bone or lesions of olfactory fibers

- ### OPTIC NERVES (CN II) (PURELY SENSORY)

  - Carry AFFERENT IMPULSES for VISION

  - **Contralateral and ipsilateral fibers arise from each retina to form the optic nerves,** pass through optic foramina of the orbits and converge to form the **OPTIC CHIASMA (or optic chiasm)**

  - They then form the optic tracts that enter the thalamus and synapse onto the LGN of the thalamus.

  - The fibers then project from thalamic nuclei as thalamic fibers (**optic radiation**) to the occipital cortex.

  - ANOPSIAS: visual defects

- Damage to CN II results in **blindness** in the eye served by the nerve.

- Damage to the visual pathway, <u>distal to the optic chiasma</u>, results in **partial visual loss**

- **OCULOMOTOR NERVES (CN III)**

  - Carry EFFERENT FIBERS, which pass through superior orbital fissure, **from ventral midbrain to the eye**

  - **Innervate 4 of 6 extrinsic eye muscles** that <u>help direct the eyeball</u> *(inferior oblique, superior / medial / inferior medial rectus muscles)* <u>and raise the upper eyelid</u> *(levator palpebrae superioris muscle)*

  - Autonomic nervous system (ANS) EFFERENTS:

    - To constrictor muscles of the iris (for <u>pupil constriction</u>) and;

    - To the ciliary muscle (<u>controls lens shape for focusing</u>)

  - Proprioceptor AFFERENTS from the 4 extrinsic muscles to the midbrain

  - OCULOMOTOR NERVE PARALYSIS: **eye cannot be moved up or inward**

    - At rest, the eye turns laterally (***external strabismus***), upper eyelid droops (***ptosis***), double vision and trouble focusing on close objects

- **TROCHLEAR NERVES (CN IV)**

  - Carry SOMATIC EFFERENTS to and PROPRIOCEPTOR AFFERENTS from the *superior oblique muscle*

  - **Fibers emerge from the dorsal midbrain and course ventrally around the midbrain** to enter the orbits of the eyes via superior orbital fissures of the sphenoid, along with CN III

  - CN IV TRAUMA or PARALYSIS: **double vision** and **reduced ability to rotate eye inferolaterally**

- **TRIGEMINAL NERVES (CN V)**

  - Carry AFFERENTS for TOUCH, TEMPERATURE and PAIN from the FACE

  - Carry BRANCHIAL EFFERENTS for CHEWING MUSCLES

- **OPHTHALMIC DIVISION (V1):** AFFERENT fibers run from the face to the **pons via superior orbital fissure of the sphenoid**; innervate skin of anterior scalp, upper eyelid and nose; afferents from nasal cavity mucosa, cornea and lacrimal gland

- **MAXILLARY DIVISION (V2):** AFFERENT fibers run from the face to the **pons via foramen rotundum of the sphenoid** bone; afferents from nasal cavity mucosa, palate, upper teeth, skin of cheek, upper lip and lower eyelid

- **MANDIBULAR DIVISION (V3):** AFFERENT fibers run from the face to pons and pass through the skull **via foramen ovale of the sphenoid** bone; afferents from anterior tongue (except taste buds), lower teeth, skin of chin and temporal region of scalp; EFFERENTS to and AFFERENTS from MUSCLES of MASTICATION (*temporalis, masseter, lateral / medial pterygoid muscles, and digastric muscles*)

- TIC DOLOUREUX (TRIGEMINAL NEURALGIA): unknown factor(s) cause CN V inflammation, but may reflect pressure on CN V root; causes symptomatic *tics*, which involves **excruciating, stabbing pain that occurs ~100x/day**

- **ABDUCENS NERVES (CN VI)**

  - Carry mainly EFFERENT fibers to, and some proprioceptor AFFERENTS from the *lateral rectus muscle* of the eye

  - **Fibers leave the inferior pons and enter the orbit of the eye via superior orbital fissure**

  - CN VI PARALYSIS: **eye cannot be moved laterally**; at rest, affected eyeball turns medially (***internal strabismus***)

- **FACIAL NERVES (CN VII)**

  - Carry MIXED fibers, which are the CHIEF MOTOR NERVES of the FACE

  - **Fibers arise from the pons**, enter temporal bone via internal acoustic meatus, and run within temporal bone before emerging through stylomastoid foramen; fibers then course to lateral aspect of the face

  - **5 major branches on the face:**
    - **TEMPORAL branch**
    - **ZYGOMATIC branch**
    - **BUCCAL branch**
    - **MANDIBULAR branch**
    - **CERVICAL branch**

- BRANCHIAL EFFERENTS to and PROPRIOCEPTOR AFFERENTS from the skeletal muscles of face for facial expression

- ANS EFFERENTS to lacrimal, nasal, palatine, submandibular and sublingual glands (some of the cell bodies are in *pterygopalatine* and *submandibular ganglia* on CN V)

- AFFERENTS from tastes buds of anterior 2/3 of the tongue and from tiny patch of skin on ear (cell bodies are in *geniculate ganglion*)

- BELL'S PALSY: **paralysis of facial muscles on affected side and partial loss of taste sensation**; caused by herpes simplex viral infection, which causes inflammation and swelling of CN VII

- **AUDITORY (or VESTIBULOCOCHLEAR) NERVES (CN VIII)** (PURELY SENSORY)

  - Carry AFFERENT IMPULSES for HEARING and EQUILIBRIUM

  - **COCHLEAR DIVISION:** carry AFFERENTS from hearing receptors located within the inner ear of temporal bone, pass through internal acoustic meatus, **enter brain stem at pons-medulla border** (cell bodies in spiral ganglion within cochlea)

  - **VESTIBULAR DIVISION:** carry AFFERENTS from equilibrium receptors (cell bodies in vestibular ganglia)

  - LESIONS of CN VIII or cochlear receptors cause **central or nerve deafness**

  - DAMAGE to VESTIBULAR DIVISION: **dizziness**, rapid involuntary eye movements, **loss of balance**, nausea and vomiting

- **GLOSSOPHARYNGEAL NERVES (CN IX)**

  - Carry MIXED fibers, which innervate part of tongue and the pharynx

  - **Fibers emerge from medulla** and leave skull via jugular foramen to run to throat

  - BRANCHIAL EFFERENTS to and PROPRIOCEPTOR AFFERENTS from the *stylopharyngeus muscle*, which elevates the pharynx during swallowing

  - ANS EFFERENTS to parotid gland

- AFFERENTS <u>conduct taste and general sensory</u> (touch, pressure, pain) <u>impulses from pharynx and posterior tongue</u>; afferents from <u>chemoreceptors</u> in carotid bodies and <u>pressure receptors</u> of carotid sinus; also innervate small area of skin on external ear

- **CN IX DAMAGE: impairs swallowing and taste** on posterior 1/3 of tongue (sour and bitter tastes)

- **VAGUS NERVES (CN X)**

  - Carry MIXED fibers that serve the pharynx, larynx, heart, lungs, abdominal viscera

  - This is the ONLY CN to EXTEND BEYOND the HEAD and NECK REGION.

  - **Fibers emerge from medulla, pass through skull via jugular foramen, and descend through neck into thorax and abdomen.**

  - BRANCHIAL EFFERENTS to skeletal muscles of pharynx and larynx, for swallowing

  - ANS EFFERENTS innervate heart, lungs and abdominal viscera (regulate heart rate, breathing and digestive activity)

  - AFFERENTS from:

    - Thoracic and abdominal viscera

    - *Carotid sinus* (mechanoreceptor for blood pressure)

    - The *carotid* and *aortic bodies* (chemoreceptors for respiration)

    - Taste buds of posterior tongue (epiglottis)

    - Mucosa of larynx and pharynx

  - Also innervate tiny area of skin on external ear and some of membrane lining middle ear

  - PROPRIOCEPTOR AFFERENTS from muscles of larynx and pharynx

  - CN X PARALYSIS: **hoarseness or loss of voice**; difficulty swallowing; impaired GI mobility

  - **TOTAL DESTRUCTION** of BOTH CN X is **FATAL** because these ANS nerves are ▪ ▪ crucial in maintaining the normal state of visceral organ activity.

- **ACCESSORY NERVES (CN XI)**

    - **CRANIAL ROOTS:** carry BRANCHIAL EFFERENTS **from lateral aspect of medulla** to larynx, pharynx and soft palate

    - **SPINAL ROOTS:** carry BRANCHIAL EFFERENTS **from superior region ($C_1 - C_5$) of spinal cord** to *trapezius* and *sternocleidomastoid muscles* (move head and neck)

    - SPINAL ROOT INJURY of ONE CN XI: **causes head to turn toward injury side** due to sternocleidomastoid paralysis; **shrugging** of the shoulder on the injured side is **difficult**

- **HYPOGLOSSAL NERVES (CN XII)**

    - Carry EFFERENTS to intrinsic and extrinsic muscles of tongue

    - **Fibers arise by a series of roots from medulla,** exit from the skull via hypoglossal canal to travel to the tongue

    - SOMATIC EFFERENTS to tongue muscles allow food mixing and manipulation by tongue during chewing; also allow tongue movements that contribute to swallowing and speech

    - CN XII DAMAGE: **difficulties in speech and swallowing; tongue deviates toward affected side;** paralyzed side eventually begins to atrophy

    - BOTH CN XII DAMAGED: **cannot protrude tongue**

Chapter 15 – THE NERVOUS SYSTEM IV:
THE AUTONOMIC DIVISION of the NERVOUS SYSTEM (ANS)

A. **INTRODUCTION to the ANS**

- The ANS is the GENERAL VISCERAL MOTOR DIVISION of the PNS.

- System of motor neurons that innervates smooth muscle, cardiac muscle, and glands

- It regulates body temperature and coordinates cardiovascular, respiratory, digestive, excretory and reproductive functions.

- Fundamentally operates at the subconscious level, to accomplish routine physiological modifications to the organ systems (**processes that maintain homeostasis**)

- **AUTONOMIC vs. SOMATIC NERVOUS SYSTEMS**

  ■ Both the ANS and SNS have **afferent** and **efferent neurons**.

  ■ But they **differ in receptor and effector organ location**.

     o In the somatic motor division of the nervous system (SNS), **a single lower motor neuron forms the pathway from the CNS to contact and exert direct control over skeletal muscles**.

     o However, in the ANS, afferent pathways originate in visceral receptors, and efferent pathways connect to visceral effector organs.

  ■ The ANS and SNS also differ in the arrangement of the neurons connecting the CNS to the effector organs.

     o **Autonomic motor pathways consist of chains of two neurons:**

     • the **PREganglionic NEURON**: a visceral motor neuron whose cell body is in the CNS, which sends its axons, called PREganglionic fibers, to synapse on

        • the **GANGLIONIC NEURON**: neuron whose cell body is in an autonomic ganglion, which sends its axons, called POSTganglionic fibers to peripheral tissues and organs, such as cardiac and smooth muscle, adipose tissue, and glands.

- ■ The <u>POST</u>ganglionic fibers of these neurons carry impulses away from the ganglion; hence, its name. Their neurons are sometimes called *POSTganglionic neurons* as well, even though their cell bodies are **within** ganglia.

- • ☐ **SUBDIVISIONS of the ANS** ☐

  - ■ The ANS has <u>parasympathetic</u> and <u>sympathetic divisions</u>, both of which innervate many of the same organs but <u>produce opposite effects</u>.

    - ○ The **PARASYMPATHETIC DIVISION** is active during the *"rest and digest"* (or *"rest and repose"*) mode.

      - • It is *craniosacral* and;

      - • Has comparatively **long PREganglionic axons (or fibers),** which synapse on;

        - ■ Neurons of **terminal ganglia** (located close to the effector organs), or;

        - ■ Neurons of **intramural ganglia** (located within the tissues of the effectors)

    - ○ The **SYMPATHETIC DIVISION** prepares the body for the *"fight or flight"* mode.

      - • It is *thoracolumbar* and;

      - • Has comparatively **long POSTganglionic axons.**

  - ■ <u>The two divisions differ in the neurotransmitter they release at the effector organ.</u>

    - ○ **ALL PREganglionic terminals release ACh** and are stimulatory.

      - • Plasmalemma receptors determine whether the response will be **stimulatory (+) or inhibitory (-).**

    - ○ **ACETYLCHOLINE** (ACh) is released by **ALL PARASYMPATHETIC POST**ganglionic **fibers**, and may be **(+)** or **(-).**

    - ○ **NOREPINEPHRINE** (NE) is released by **MOST SYMPATHETIC POST**ganglionic **fibers**; effects are usually **(+).**

B. THE PARASYMPATHETIC DIVISION

- BASIC ORGANIZATION

  - Consists of PREganglionic neurons in the brain stem and in the lateral portion of the anterior gray horns of S_2-S_4; their fibers leave the brain in CN III (oculomotor), VII (facial), IX (glossopharyngeal), and X (vagus).

  - **PREganglionic fibers in CN III, VII, and IX** innervate visceral structures in the <u>head</u>, and they **synapse in the ciliary, pterygopalatine, submandibular, and otic ganglia**.

  - **Short POSTganglionic fibers** then connect to their peripheral target organs or tissues.

  - **PREganglionic fibers in CN X synapse in intramural ganglia** within structures in the <u>thoracic cavity</u> and in the <u>abdominopelvic cavity</u> (as far as the last segments of the large intestine).

  - **The VAGUS NERVE (CN X) provides roughly 75% of all parasympathetic outflow.**

  - The **sacral outflow** does NOT join the ventral rami of the spinal nerves.

    - The **PREganglionic fibers form distinct PELVIC NERVES**, which innervate INTRAMURAL GANGLIA in the <u>kidney</u> and <u>urinary bladder</u>, the terminal portions of the <u>large intestine</u>, and the <u>sex organs</u>.

  - PREganglionic fibers release ACh and stimulate ganglionic neurons.

  - **ALL POSTganglionic fibers release ACh** at neuroeffector junctions.

- GENERAL FUNCTIONS

Major parasympathetic effects center on relaxation, food processing and energy absorption:

- Pupil constriction

- Digestive gland secretion, including the salivary glands, gastric glands, duodenal and other intestinal glands, the pancreas, and the liver.

- Hormone secretion for nutrient absorption

- Increased digestive tract activity

- Defecation activities

- Urinary bladder contraction

- Respiratory passageway constriction

- Decreased heart rate

- Sexual arousal

- **ACTIVATION and NEUROTRANSMITTER RELEASE**

  - All parasympathetic neurons are **cholinergic**.

    - i.e., All parasympathetic PREganglionic and POSTganglionic fibers **release ACh** at synapses and neuroeffector junctions.

  - **Effects are short-lived** because of the actions of enzymes at the postsynaptic plasmalemma and in the surrounding tissues.

  - **Two different types of ACh receptors in POSTsynaptic plasmalemmae:**

    - **NICOTINIC** receptors – located on ganglionic cells of both divisions of the ANS, and at neuromuscular synapses

      - Exposure to ACh causes excitation by opening plasmalemmae channels

    - **MUSCARINIC** receptors – located at neuroeffector junctions in the parasympathetic division and those cholinergic neuroeffector junctions in the sympathetic division

      - Stimulation of muscarinic receptors produces a longer-lasting effect than does stimulation of nicotinic receptors

- **CRANIAL OUTFLOW**

  - Cranial parasympathetic fibers **arise in the brain stem nuclei of CN III, VII, IX, and X**; and they synapse in ganglia in the head, thorax, and abdomen.

- o Fibers in **CN VII** serve the submandibular, sublingual, lacrimal, and nasal glands; and synapse in the **SUBMANDIBULAR** and **PTERYGOPALATINE GANGLIA** (singular = *ganglion*).

- o Fibers in **CN IX** serve the parotid gland and synapse in the **OTIC GANGLION**.

- Parasympathetic fibers in the vagus nerve (**CN X**) innervate organs in the thorax and most of the abdomen, including the heart, lungs, esophagus, stomach, liver, and most of the intestines.

  - o Fibers in the **vagus nerve** are **PREganglionic**.

  - o **Almost all ganglionic neurons** are located in **INTRAMURAL GANGLIA** within the organ walls.

- · | **SACRAL OUTFLOW** |

  - Sacral parasympathetic pathways **innervate the pelvic viscera.**

  - The **PREganglionic fibers** exit from the visceral motor region of the gray matter of the spinal cord (S_2-S_4) and **form the PELVIC SPLANCHNIC NERVES.**

  - **Most of these fibers synapse in intramural ganglia** in the kidney, bladder, latter portions of the large intestine, and sex organs.

- · | **SUMMARY** |

  - **Visceral motor nuclei** of the parasympathetic division are associated with **CN III, VII, IX, and X**, and with **sacral segments S_2-S_4.**

  - **Ganglionic neurons** are located in **terminal or intramural ganglia** near or within target organs, respectively.

  - Parasympathetic fibers innervate areas controlled by cranial nerves and organs in the thoracic and abdominopelvic cavities.

  - **All parasympathetic neurons are cholinergic.**

  - Parasympathetic **effects** are usually **short-lived** and **restricted** to specific target areas.

C. THE SYMPATHETIC DIVISION

- BASIC ORGANIZATION

  - The sympathetic division consists of:

    - PREganglionic neurons between T_1 and L_2

    - **Ganglionic neurons in sympathetic CHAIN GANGLIA** (which innervate effectors in the body wall, head and neck, limbs and inside the thoracic cavity) **and COLLATERAL GANGLIA** (which innervate effectors in the abdominopelvic cavity)

    - **Specialized neurons within the suprarenal glands**

  - The **PREganglionic sympathetic cell bodies** are in the **lateral horn** of the spinal gray matter from the level of T_1 **to** L_2.

  - The sympathetic division **supplies some peripheral structures** that the parasympathetic division does not: arrector pili, sweat glands, and the smooth muscle of blood vessels.

  - Sympathetic ganglia include 22-24 pairs of:

    - **Sympathetic TRUNK GANGLIA** (also called **CHAIN GANGLIA** and **PARAVERTEBRAL GANGLIA**), which are linked together to form sympathetic trunks on both sides of the vertebral column

    - Plus one **unpaired COLLATERAL GANGLIA** (also called **PREVERTEBRAL GANGLIA**), most of which lie on the aorta in the abdomen, located anterior to the vertebral column

  - PREganglionic fibers are cholinergic -- release ACh (excitatory) -- and stimulate ganglionic neurons.

  - Most POSTganglionic fibers release NE at neuroeffector junctions, causing a "Fight or Flight" response.

- SYMPATHETIC PATHWAYS

  - ■ SYMPATHETIC CHAIN GANGLIA

    - o **Every PREganglionic sympathetic fiber** leaves the <u>lateral gray horn of the thoracolumbar</u> spinal cord through a **ventral root** and **spinal nerve**.

      - From there, it runs in a **WHITE RAMUS COMMUNICANS (myelinated)** to a sympathetic trunk ganglion or collateral ganglion, where it synapses with the ganglionic neuron that extends to the visceral effector.

      - **Many PREganglionic axons ascend or descend in the sympathetic trunk to synapse in a ganglion at another body level.**

    - o **In each sympathetic chain**, there are 3 cervical, 11-12 thoracic, 2-5 lumbar, and 4-5 sacral ganglia, and 1 coccygeal sympathetic ganglion

      - Every spinal nerve has a **GRAY RAMUS** that <u>carries</u> sympathetic **POSTganglionic fibers**.

      - Only **thoracic** and **superior lumbar ganglia** receive **PREganglionic fibers** via **WHITE RAMI**.

      - The **cervical, inferior lumbar,** and **sacral chain ganglia** receive **PREganglionic** innervation **from collateral fibers** of sympathetic neurons.

      - Every spinal nerve <u>receives</u> a **GRAY RAMUS** from a ganglion of the sympathetic chain.

    - o <u>In the sympathetic pathway to the **BODY PERIPHERY**</u> (to innervate arrector pili, sweat glands, and peripheral blood vessels), the **PREganglionic fibers** synapse in the **sympathetic trunk ganglia** and the postganglionic fibers run in **GRAY RAMI COMMUNICANTES (unmyelinated)** to the **DORSAL** and **VENTRAL RAMI** of the spinal nerves for peripheral distribution.

    - o <u>In the sympathetic pathway to the **HEAD**</u>, the **PREganglionic fibers** synapse in the **SUPERIOR CERVICAL GANGLION**. From there, **most POSTganglionic fibers associate with a large artery** that distributes them to the glands and smooth musculature of the head.

    - o <u>In the sympathetic pathway to **THORACIC ORGANS**</u>, **most PREganglionic fibers synapse in the nearest SYMPATHETIC TRUNK GANGLION**, and the **POSTganglionic fibers run directly to the organs** (lungs, esophagus).

- Many POSTganglionic fibers to the heart, however, descend from the cervical ganglia in the neck.

- ▪ | **COLLATERAL GANGLIA** |

  - ○ In the sympathetic pathway to **ABDOMINAL ORGANS**, the **PREganglionic fibers** run in **splanchnic nerves (*greater, lesser, lumbar,* and *sacral*)** to synapse in **collateral ganglia** on the aorta.

    - From these ganglia, the POSTganglionic fibers follow large arteries to the abdominal viscera (stomach, liver, kidney, and most of the large intestine).

  - ○ In the sympathetic pathway to **PELVIC ORGANS**, the **PREganglionic fibers** synapse in **sympathetic trunk ganglia or in collateral ganglia** on the aorta, sacrum, and pelvic floor.

    - POSTganglionic fibers travel through the most inferior autonomic plexuses to the pelvic organs.

  - ○ The splanchnic nerves innervate the hypogastric plexus and three COLLATERAL GANGLIA:

    - The **CELIAC ganglion** – innervates the stomach, liver, pancreas, and spleen

    - The **SUPERIOR MESENTERIC ganglion** – innervates the small intestine and initial segments of the large intestine

    - The **INFERIOR MESENTERIC ganglion** – innervates the kidney, bladder, sex organs, and terminal segments of the large intestine

- | **The ROLE of the ADRENAL MEDULLA in the SYMPATHETIC DIVISION** |

  - ▪ The **adrenal glands**, individually located superior to each kidney, contain a **MEDULLA of modified POSTganglionic sympathetic neurons** that secrete the hormones **epinephrine (E, or adrenaline)** and **norepinephrine (NE)** into the blood.

    - ○ The result is the **"surge of adrenaline"** felt during excitement.

    - ○ Some PREganglionic fibers do NOT synapse as they pass through both the sympathetic chain ganglia and collateral ganglia. They enter one of the suprarenal glands instead and synapse on modified neurons within the suprarenal medulla.

- The cells of the adrenal medulla are **innervated by PREganglionic sympathetic neurons**, which signal them to secrete the hormones, E and NE into the circulation, causing a prolonged sympathetic innervation effect.

- ## PLASMALEMMA RECEPTORS and NEUROTRANSMITTERS

  - There are two classes of sympathetic receptors, which are stimulated by NE and E:

    - **ALPHA receptors** – respond to stimulation by depolarizing the plasmalemma

    - **BETA receptors** – respond to stimulation by changing the metabolic activity of the cells

  - **Most POSTganglionic fibers are adrenergic (release NE)**, but a **few are cholinergic (release ACh)**. POSTganglionic fibers that <u>innervate sweat glands</u> of the skin and <u>blood vessels</u> to skeletal muscles are **cholinergic**.

- ## SUMMARY of SYMPATHETIC CHARACTERISTICS

  - The sympathetic division consists of **2** segmentally arranged **sympathetic chains** that lie **lateral** to the vertebral column; **3 collateral ganglia** that lie **anterior** to the vertebral column; and **2 suprarenal medullae**.

  - **PREganglionic fibers** are relatively **short**, *except* for those of the suprarenal medulla.

  - **POSTganglionic fibers** are quite **long**.

  - **Extensive divergence** typically occurs.

    - A single PREganglionic fiber synapses with many ganglionic neurons in different ganglia.

  - **All PREganglionic fibers** are **cholinergic** (release ACh).

  - **Most POSTganglionic fibers** are **adrenergic** (release NE).

  - **Effector response depends on the nature and activity of the receptor.**

- In a crisis, SYMPATHETIC ACTIVATION occurs, in which the entire sympathetic division responds.

- Its effects include increased alertness, a feeling of energy and euphoria, increased cardiovascular and respiratory activities, general elevation in muscle tone, and mobilization of energy reserves.

D. DUAL INNERVATION

- Organs with DUAL INNERVATION **receive instructions from both the parasympathetic and sympathetic divisions** of the ANS.

- **Nerves from both divisions** intermingle (in body cavities) to **form nerve plexuses** (nerve networks), which include the **cardiac, pulmonary, esophageal, celiac, inferior mesenteric,** and **hypogastric** plexuses.

E. CENTRAL CONTROL OF THE ANS

- VISCERAL MOTOR FUNCTIONS are influenced by the **medulla oblongata**, the periaqueductal gray matter, spinal visceral reflexes, the **hypothalamus**, the **amygdala**, and the **cerebral cortex.**

  - **Visceral reflexes** are the simplest functions of the ANS, which **provide automatic motor responses** that can be **modified, facilitated, or inhibited** by higher centers, especially in the **hypothalamus.**

  - Higher brain centers in the **posterior and lateral hypothalamus:** involved in coordination and regulation of sympathetic function

  - Portions of the **anterior and medial hypothalamus:** involved in the control of parasympathetic functioning

- Some people can voluntarily regulate some autonomic activities by gaining extraordinary control over their emotions.

Chapter 16 – THE NERVOUS SYSTEM V:
THE GENERAL and SPECIAL SENSES

A. **VISUAL SENSORY NEURONS**

- General visceral sensory neurons **monitor temperature, pain, irritation, chemical changes, and stretch** in the visceral organs.

- **PERIPHERAL SENSORY RECEPTORS** : pick up stimuli (environmental changes) from inside and outside the body, then initiate impulses in sensory axons

  - **CLASSIFIED by LOCATION of the STIMULUS:**

    o EXTEROCEPTORS: sensitive to stimuli arising outside the body

    o INTEROCEPTORS (*visceroceptors*): sensitive to stimuli from internal viscera

    o PROPRIOCEPTORS: body movements sensory information

  - **CLASSIFIED by STIMULUS DETECTED:**

    o **MECHANORECEPTORS:** respond to mechanical forces such as touch, pressure, stretch and vibrations

      - **TACTILE RECEPTORS** – detect touch, pressure and vibration

      - **BARORECEPTORS** (*stretch receptors*) – detect pressure changes in the walls of blood vessels and the walls of the digestive, reproductive and urinary tracts

      - **PROPRIOCEPTORS** – muscle spindles that respond to the positions of the joints, the tension in tendons and ligaments, and the state of muscular contraction

    o **THERMORECEPTORS:** respond to temperature changes; also conduct sensations along the same pathways that carry pain sensation

    o **CHEMORECEPTORS:** respond to chemicals in solution and changes in blood chemistry

    o **PHOTORECEPTORS:** sensitive to light

○ **NOCICEPTORS:** respond to harmful stimuli (usually associated with tissue damage) that result in pain

- FAST PAIN – prickling sensation

- SLOW PAIN – burning and aching sensation

- **CLASSIFIED by STRUCTURE:**

  ○ FREE (naked) DENDRITIC ENDINGS: respond chiefly to PAIN and TEMPERATURE

    - **MERKEL DISC (or TACTILE DISC):** disc-shaped epithelial cell innervated by a dendrite (**slowly adapting** receptor)

    - **ROOT HAIR PLEXUSES** (singular = *plexus*): receptors for <u>light touch</u> that monitor bending of hairs (**rapidly adapting** receptor)

  ○ ENCAPSULATED DENDRITIC ENDINGS: all seem to be MECHANORECEP-TORS; consist of one or more end fibers of sensory neurons enclosed by a capsule of connective tissue; capsule serves to amplify stimulus or filter out wrong types of stimuli

    - **MEISSNER'S CORPUSCLES (or TACTILE CORPUSCLES):** sensitive to fine, discriminative touch

    - **KRAUSE'S END BULBS:** a type of Meissner's corpuscle for fine touch

    - **PACINIAN'S CORPUSCLES (or LAMELLATED CORPUSCLES):** respond to deep pressure, specifically vibrations

    - **RUFFINI'S CORPUSCLES:** respond to pressure and light touch

    - **PROPRIOCEPTORS:** sensitive to stretch in locomotory organs

      - <u>Muscle spindles</u>: respond to changes in length of a muscle

      - <u>Golgi tendon organs</u>: monitor tension within tendons

      - <u>Joint kinesthetic receptors</u>: monitor stretch in synovial joints

- Visceral sensory fibers run within the autonomic nerves, especially within the vagus and the sympathetic nerves. The sympathetic nerves **carry most pain fibers from the visceral organs of the body trunk**.

- A <u>simplified description of most visceral sensory pathways</u> to the brain is the following:

  - **SENSORY NEURONS** to **SPINOTHALAMIC TRACT** to **THALAMUS** to **VISCERAL SENSORY CORTEX**

- **Visceral pain** is induced by stretching, infection, and cramping of internal organs but seldom by cutting or scraping these organs.

  - Pain in visceral organs is **referred to somatic regions of the body** that receive innervation from the same spinal cord segments.

- **Many visceral reflexes are spinal reflexes**, such as the defecation reflex.

  - Some visceral reflexes, however, involve only **peripheral neurons**.

B. | **INTRODUCTION to THE SPECIAL SENSES** |

- **The SPECIAL SENSES: olfaction (smell), gustation (taste), equilibrium, hearing and vision**

- **Receptors** for the special senses are located in specialized areas, or sense organs.

- **SENSORY RECEPTOR:** a specialized cell that sends a sensation to the CNS, when stimulated

- **RECEPTOR SPECIFICITY:** allows each receptor to respond to particular stimuli

  - FREE NERVE ENDINGS are the simplest form of receptors.

  - RECEPTIVE FIELD – the area monitored by a single receptor cell

- **INTERPRETATION of SENSORY INFORMATION**

  - **TONIC receptors** – constantly sending signals to the CNS

- **PHASIC receptors** – become active only when the conditions (that they monitor) change

- **CENTRAL PROCESSING and ADAPTATION**

  - **ADAPTATION** – reduction in sensitivity in the presence of a constant stimulus

    o **PERIPHERAL (or sensory)** adaptation – involves changes in receptor sensitivity

    o **CENTRAL** adaptation – inhibition along the sensory pathways occur

  - **FAST-ADAPTING** receptors are PHASIC.

  - **SLOW-ADAPTING** receptors are TONIC.

- **SENSORY LIMITATIONS** - incomplete information provided by our sensory receptors is due to the following reasons:

  - We do not have receptors for every stimulus.

  - Limited ranges of sensitivity of our receptors

  - CNS interpretation of a stimulus is filtered and limited.

C. **THE CHEMICAL SENSES** : TASTE and SMELL

- **TASTE (or GUSTATION)** provides information about the food and liquids that are consumed

  - Gustatory receptors are clustered in taste buds, each of which contains gustatory cells that project *taste hairs* through a narrow taste pore.

  - Most taste buds are on the tongue, in the epithelium of **fungiform** and **circumvallate papillae**.

  - Taste buds contain **gustatory (taste) cells** and **basal cells** that replace damaged gustatory cells. The gustatory cells are excited when taste-stimulating chemicals bind to their microvilli.

  - The five basic qualities of taste are **sweet, sour, salty, bitter, and umami**.

- The sense of taste is served by **cranial nerves VII, IX, and X**, which send impulses to the medulla. From there, impulses travel to the thalamus and the taste area of the cerebral cortex.

- ### SMELL (or OLFACTION)

  - The **olfactory epithelium** is located in the roof of the nasal cavity.

    - This epithelium **contains olfactory receptors, supporting cells, and basal cells**.

  - The olfactory receptor cells are **modified ciliated bipolar neurons, which are sensitive to chemicals dissolved in the overlying mucus**.

    - Odor molecules **bind to the cilia**, exciting the neurons.

    - **Axons** of these receptor neurons **form the filaments** of the olfactory nerve (CN I).

  - Olfactory nerve axons transmit impulses to the **olfactory bulb**.

    - Here, these axons synapse with **mitral cells** in structures called **glomeruli**.

  - After receiving input from the olfactory receptor neurons, the mitral cells send this olfactory information through the olfactory tract to the olfactory cortex and limbic system.

- ### DISORDERS of the CHEMICAL SENSES

  - Disorders of smell include *anosmia* (inability to smell) and *uncinate fits* (smell hallucinations).

D. ### THE EYE AND VISION

- The eye is located in the bony orbit and is cushioned by fat.

- The cone-shaped orbit also contains nerves, vessels, and extrinsic muscles of the eye.

- ### ACCESSORY STRUCTURES of the EYE

  - <u>EYEBROWS</u> shade and protect the eyes.

- **EYELIDS (or *palpebrae*)** protect and lubricate the eyes by reflexive blinking.

  o The free margins of the upper and lower eyelids are separated by the **palpebral fissure**, but both are connected at the **MEDIAL CANTHUS** and the **LATERAL CANTHUS**.

  o The **CONJUNCTIVA** is the epithelium covering the inner surface of the eyelids and the outer surface of the eye. Its mucus lubricates the eye surface

    • **PALPEBRAL conjunctiva** – covers the inner surface of the eyelids

    • **OCULAR conjunctiva (or BULBAR conjunctiva)** – covers the anterior surface of the eye

  o Each eyelid contains a supporting **TARSAL PLATE** (broad sheets of connective tissue), the roots of the **eyelashes**, and **tarsal** and **ciliary glands**.

    • Each **eyelash** is monitored by a **root hair plexus**, and displacement of the hair triggers a **blinking reflex**.

    • Large sebaceous glands, called the *glands of Zeis*, are associated with the eyelashes.

  o MUSCLES in the eyelids include the **levator palpebrae superioris** (which opens the eye) and the **orbicularis oculi** (which closes the eye).

  o **TARSAL GLANDS (or *Meibomian glands*)**, that line the **palpebral margin** of the eyelids, secrete a lipid-rich product that prevents the eyelids from sticking together.

  o **GLANDS** within the **LACRIMAL CARUNCLE**, at the medial canthus, produce the thick secretions that contribute to the gritty deposits occasionally found after a good night's sleep.

- The **LACRIMAL APPARATUS** produces, distributes, and removes tears. It consists of the following:

  o The **lacrimal gland** (tear gland on dorsolateral surface of the eye) secretes lacrimal fluid (tears), which is blinked medially across the eye surface and drained into the nasal cavity through lacrimal canaliculi.

    • Lacrimal fluid (tears) is slightly alkaline and contains **lysozymes** (antibacterial enzymes).

- Tears collect in the *lacus lacrimalis*, and pass through the *lacrimal punta* (small holes in the surface of the lacrimal bone) before reaching the lacrimal canaliculi.

  o **Superior** and **inferior lacrimal canaliculi** – two small grooves in the surface of the lacrimal bone

  o **Lacrimal sac** – collects tears from the lacrimal canaliculi

  o **Nasolacrimal duct** – extends along the nasolacrimal canal formed by the lacrimal bone and the maxilla to deliver the tears to the inferior meatus on the same side of the nasal cavity

- The six ╎ EXTRINSIC EYE MUSCLES ╎ are:

  o The **LATERAL** and **MEDIAL RECTUS** turn the eye laterally and medially, respectively.

  o The **SUPERIOR** and **INFERIOR RECTUS** elevate and depress the eye, respectively, but also turn it medially.

  o The **SUPERIOR** and **INFERIOR OBLIQUES** depress and elevate the eye, respectively, but also turn it laterally.

- ╎ **ANATOMY of the EYEBALL** ╎

  - The WALL of the eye has three layers.

    o The most external, ╎ **FIBROUS TUNIC or LAYER** ╎ consists of the **POSTERIOR SCLERA** and the **CORNEAL LIMBUS**.

      - The tough **SCLERA** (dense, fibrous connective tissue of the fibrous tunic) protects the eye and gives it shape.

      - The **CORNEAL LIMBUS** is the clear window through which light enters the eye.

    o The middle, pigmented ╎ **VASCULAR TUNIC** ╎ consists of the choroid, the ciliary body, and the iris.

- The **CHOROID** provides nutrients to the retina's photoreceptors and prevents the scattering of light within the eye.

- The **CILIARY BODY** contains smooth ciliary muscles that control the shape of the lens and ciliary processes that secrete aqueous humor, and attach to the suspensory ligaments (ZONULAR FIBERS) of the lens.

- The **IRIS** contains smooth muscle that changes the size of the pupil, which forms the boundary between the anterior and posterior chambers.

o The innermost │ **NEURAL (sensory) TUNIC or LAYER** │ contains the neural **retina** and the **optic nerve**.

- The **RETINA** consists of an *outer pigmented layer* and an *inner neural layer*, which contains visual receptors and associated neurons.

- The **NEURAL LAYER** contains **photoreceptors (rod and cone cells)** and other types of neurons.

  - **RODS** provide black-and-white vision in dim light.

  - **CONES** provide color vision in bright light.

- Light stimulates the photoreceptors, which signal **BIPOLAR CELLS**, then send signals to **GANGLION CELLS**, which transmit signals to the brain via the optic nerve.

  - **HORIZONTAL cells** and **AMACRINE cells** modify the signals passed between other retinal components.

- The axons of ganglion neurons run along the inner retinal surface toward the optic disc, forming the optic nerve.

o The outer segments of the rods and cones contain **light-absorbing pigment** in membrane-covered discs. Light modifies this pigment to initiate the flow of signals through the visual pathway.

- Two important spots on the POSTERIOR RETINAL WALL are :

  ○ The **MACULA LUTEA** (concentrated with cones) with its **FOVEA CENTRALIS** (area of highest visual acuity) and

  ○ The **OPTIC DISC** (blind spot), where axons of ganglion cells form the optic nerve.

- The outer third of the retina (photoreceptors) is nourished by capillaries in the choroid, whereas the inner two-thirds is supplied by the central vessels of the retina.

- The **POSTERIOR SEGMENT** of the eye , posterior to the lens, is called the **VITREOUS CHAMBER**, which contains the gel-like **vitreous humor (or vitreous body)**.

  ○ The **VITREOUS HUMOR or BODY** helps <u>stabilize the shape</u> of the eye and <u>support</u> the retina.

- The **ANTERIOR SEGMENT** of the eye , anterior to the lens, is divided into **ANTERIOR and POSTERIOR CHAMBERS** by the iris.

  ○ The anterior segment is filled with **AQUEOUS HUMOR**.

  ○ Aqueous humor continually forms at the CILIARY PROCESSES in the posterior chamber, flows into the anterior chamber, and drains into the **SCLERAL VENOUS SINUS (or CANAL of SCHLEMM)**, then reenters the circulation.

- The **BICONVEX LENS** helps to focus light and focuses a visual image on the retinal receptors.

  ○ It is suspended in the eye by the **CILIARY ZONULE** (suspensory ligaments) attached to the ciliary body.

  ○ The lens also lies posterior to the cornea and forms the anterior boundary of the vitreous chamber.

  ○ Tension in the ciliary zonule resists the lens's natural tendency to round up.

- **THE EYE as an OPTICAL DEVICE**

  - As it enters the eye, **light is bent by the cornea and the lens and focused on the retina**.

o The cornea accounts for most of this refraction, but the lens allows focusing on objects at different distances.

- **The resting eye is set for distance vision.**

  o Focusing on near objects requires **ACCOMMODATION** (allowing the lens to round as ciliary muscles release tension on the ciliary zonule).

  o The pupils also constrict.

  o Both these actions are controlled by **parasympathetic fibers** in the oculomotor nerve.

- Eye-focusing disorders include:

  o **MYOPIA** (nearsightedness),

  o **HYPEROPIA** (farsightedness),

  o **PRESBYOPIA** (loss of lens elasticity with age), and;

  o **ASTIGMATISM** (caused by unequal curvatures in different parts of the cornea or lens).

- **VISUAL PATHWAYS**

  - The visual pathway to the brain begins with some processing of visual information in the retina.

    o From there, **ganglion cell axons** carry impulses via the optic nerve, optic chiasma, and optic tract to the **lateral geniculate nucleus of the thalamus.**

    o Thalamic neurons project to the **primary visual cortex.**

  - At the **optic chiasma**, axons from the medial halves of the retinas **decussate**.

    o This phenomenon provides each visual cortex with information on the opposite half of the visual field as seen by both eyes.

    o The visual cortex compares the views from the two eyes and generates **depth perception**.

- Visual inputs to the **suprachiasmatic nucleus** and the **pineal gland** affect the function of other brainstem nuclei, which establish a visceral circadian rhythm that is tied to the day-night cycle and affects other metabolic processes.

- ### DISORDERS of the EYE and VISION

  - Blinding disorders can result from damage to the retina (**age-related macular degeneration** and **retinopathy of prematurity**) and from damage to the cornea (**trachoma**).

E. ### THE EAR: HEARING and EQUILIBRIUM

- ### THE EXTERNAL EAR

  - The **auricle** and **external acoustic meatus** constitute the external ear. The auricle surrounds the external acoustic meatus, which ends at the **tympanic membrane**.

  - The external ear functions in gathering sound waves.

  - The **tympanic membrane, or tympanum,** (eardrum) transmits sound vibrations to the middle ear.

- ### THE MIDDLE EAR

  - The middle ear is a small **tympanic cavity** within the temporal bone.

    o Its boundaries are the eardrum laterally, the bony wall of the inner ear medially, a bony roof, a thin bony floor, a posterior wall that opens into the mastoid antrum, and an anterior wall that opens into the pharyngotympanic tube.

    o It encloses and protects the **auditory ossicles**, which connect the tympanic membrane with the receptor complex of the inner ear.

  - The **pharyngotympanic, or auditory, tube**, which consists of bone and cartilage, communicates with the nasopharynx and **equalizes air pressure** across the eardrum.

  - The **auditory ossicles** (MALLEUS, INCUS, and STAPES):

    o Help to amplify sound

    o Span the middle ear cavity

- o Transmit sound vibrations from the eardrum to the oval window

- o The tiny **tensor tympani and stapedius muscles** dampen the vibrations of very loud sounds.

  - • These **muscles contract to reduce the amount of motion of the tympanum** when very loud sounds arrive.

- • **THE INNER EAR**

  - ▪ The inner ear contains the sensory organs for the perception of **equilibrium** (*bony labyrinth*) and **hearing** (*membranous labyrinth*).

  - ▪ It consists of the **BONY LABYRINTH** (*semicircular canals, vestibule, and cochlea*), which is a chamber that contains the **MEMBRANOUS LABYRINTH** (*semicircular ducts, utricle and saccule, and cochlear duct*).

    - o The bony labyrinth contains **perilymph**, whereas the membranous labyrinth contains **endolymph**.

  - ▪ The vestibule includes a pair of membranous sacs, the **SACCULE** and the **UTRICLE**, each of which contains a **MACULA**, a spot of receptor epithelium that monitors static equilibrium (sensations of gravity) and linear acceleration.

    - o The saccule and utricle are connected by a passageway that is continuous with the **endolymphatic duct**, which terminates in the **endolymphatic sac**.

    - o A macula contains clusters of **hair cells**, whose **stereocilia** ("hairs") are anchored in an overlying **otolithic membrane**.

    - o The hair cells' stereocilia contact **OTOLITHS**, which consist of densely packed mineral crystals called **statoconia**, in a gelatinous matrix.

    - o **Forces on the otolithic membrane**, caused by gravity and linear acceleration of the head, **bend the hairs and initiate impulses** in the **vestibular nerve**.

  - ▪ The **SEMICIRCULAR DUCTS** lie in three planes of space (anterior vertical, posterior vertical, and lateral), and are continuous with the utricle.

    - o The semicircular ducts provide information about the direction and strength of varied mechanical stimuli.

    - o Each semicircular duct's **CRISTAE AMPULLARES**, or **ampullae, contain hair cells** (sensory receptors) that monitor rotational acceleration.

o The "hairs" of these cells are anchored in an overlying gelatinous **CUPULA**.

o **Forces on the cupula**, caused by rotational acceleration of the head, **bend the hairs and initiate impulses** in the **vestibular nerve**.

- The **coiled COCHLEA** is divided into three parts (*scalae*).

   o The **SCALA MEDIA (or COCHLEAR DUCT)**, which contains the spiral organ, runs through its center.
      • It is an elongated portion of the membranous labyrinth.

   o The **SCALA VESTIBULI** and the **SCALA TYMPANI** are parts of the bony labyrinth.

- ▢ **MECHANISM of HEARING**

   o Sound vibrations are transmitted toward the **tympanum**; the **auditory ossicles** conduct vibrations to the base of the **stapes**, which vibrate the fluids in the **cochlea's vestibular duct**, and then passed on to the **tympanic duct**.

   o The pressure waves distort the basilar membrane and spiral organ (**organ of Corti**).

   o In turn, the **hairs of the receptor cells (hair cells of the spiral organ)**, whose tips are anchored in a nonmoving **tectorial membrane**, are bent.

   o Bending of the hairs produces impulses in the **cochlear nerve**.

- ▢ **AUDITORY and EQUILIBRIUM PATHWAYS**

   - Impulses generated by the **equilibrium receptors** travel along the **vestibular nerve** to the **vestibular nuclei** and the **cerebellum**.

      o These brain centers initiate responses that **maintain balance**.

      o There is also a minor equilibrium pathway to the posterior insula of the cerebral cortex.

   - Impulses generated by the **hearing receptors** (*sensory neurons for hearing located in the spiral ganglion of the cochlea*) travel along the **cochlear nerve** to the **cochlear nuclei in the medulla**. (Afferent fibers form the *cochlear branch of CN VIII*, synapsing at the *cochlear nucleus*.)

      o From there, auditory information passes through several nuclei in the brain stem (superior olivary, inferior colliculus) to the **medial geniculate nucleus** of the thalamus and to the **auditory cortex**.

- ## DISORDERS of EQUILIBRIUM and HEARING

  - **Motion sickness**, brought on by particular movements, cause nausea and vomiting.

  - **Ménière's syndrome** is an overstimulaiton of the hearing and equilibrium receptors caused by an excess of endolymph in the membranous labyrinth.

  - **Conduction deafness** results form interference with the conduction of sound vibrations to the internal ear.

  - **Sensorineural deafness** reflects damage to auditory receptor cells or neural pathways.

Chapter 17 – THE ENDOCRINE SYSTEM

INTRODUCTION

- The nervous system and the endocrine system function together to **monitor and adjust physiological activities for the regulation of** *homeostasis* (the maintenance of a relatively constant internal environment).

    - The <u>regulatory effects of the endocrine system</u> are **long-term ongoing metabolic processes**, compared to the short-term effects of the nervous system.

- Endocrine organs are **ductless glands that release hormones** into the blood or lymph.

- **Hormones** are messenger molecules that travel in the circulatory vessels and signal physiological changes in target cells.

- **Hormonally regulated processes** include reproduction, growth, mobilization of body defenses against stress, maintenance of the proper chemistry of the blood and body fluids, and regulation of cellular metabolism.

THE ENDOCRINE SYSTEM: AN OVERVIEW

- ### ENDOCRINE ORGANS

    - The endocrine organs are small and widely separated from one another within the body.

        o The **pure endocrine organs** are the **pituitary, thyroid, parathyroid, adrenal,** and **pineal glands.**

        o Other organs that contain endocrine cells are the **gonads, pancreas, kidney, alimentary canal, heart, thymus,** and **skin.**

        o The **hypothalamus** of the brain is a *neuroendocrine organ.*

    - Endocrine organs are **richly vascularized.**

    - Although most endocrine cells are **modified epithelial cells**, others are **neurons, muscle cells,** or **fibroblast-like cells.**

- ## HORMONES

  - Most hormones are **either amino acid derivatives** (amines, peptides, proteins) **or steroids** (lipid-based molecules derived from cholesterol).

  - Hormones produce their effects by leaving the capillaries and <u>binding to specific receptor molecules in or on their target cells</u>.

    - o Such binding triggers a **preprogrammed response in the target cell**.

  - Endocrine organs are stimulated to release their hormones by **humoral, neural, or hormonal stimuli**.

    - o Hormonal secretion is **controlled by feedback loops**.

  - <u>The hypothalamus of the brain regulates many functions of the endocrine system through the hormones it secretes.</u>

THE MAJOR ENDOCRINE ORGANS

- ## THE PITUITARY GLAND

  - The golf club-shaped pituitary gland is **suspended from the diencephalon of the brain by its stalk (INFUNDIBULUM)** and lies in the hypophyseal fossa of the sella turcica of the sphenoid bone.

  - It consists of an anterior adenohypophysis and a posterior neurohypophysis.

  - The **anterior** **ADENOHYPOPHYSIS** has three parts: (*pars distalis, pars intermedia,* and *pars tuberalis*)

    - o **PARS DISTALIS** (anterior lobe): the largest, anteriormost part of the adenohypophysis

      - Cells in the pars distalis cluster into **spheres (and branching cords)**.

      - Its <u>five cell types</u> secrete seven protein hormones (in parentheses):
        1. **SOMATOTROPIC cells** (secrete *growth hormone*, **GH**): the most abundant cells in the pars distalis

2. **MAMMOTROPIC** cells (secrete *prolactin*, **PRL**)

3. **THYROTROPIC** cells (*thyroid-stimulating hormone*, **TSH**)

4. **CORTICOTROPIC** cells (*adrenocorticotropic hormone* and *melanocyte-stimulating hormone*, **ACTH and MSH**)

5. **GONADOTROPIC** cells (*follicle-stimulating hormone* and *luteinizing hormone*, **FSH and LH**)

- The BASIC FUNCTIONS of each adenohypophyseal hormone are as follows:
  - **GH** (or *somatotropic hormone*, SH; or *somatotropin*) **stimulates growth of the body and skeleton.**

  - **PRL** signals **milk production.**

  - **TSH** signals the thyroid gland to **secrete thyroid hormone.**

  - **ACTH** signals the adrenal cortex to **secrete glucocorticoids; MSH causes the skin to darken.**

  - **FSH and LH** (together, they are called *gonadotropins*) signal the **maturation of sex cells** and the **secretion of sex hormones.**

  - Four of these seven hormones (**FSH, LH, ACTH,** and **TSH**) **stimulate other endocrine glands to secrete** and are called **tropic hormones.**

  o **PARS INTERMEDIA:** located just posterior to pars distalis

  o **PARS TUBERALIS:** located just superior to pars intermedia, and wraps around the infundibulum like a tube

- The **posterior** | **NEUROHYPOPHYSIS** | also has three parts:

  o **PARS NERVOSA** (posterior lobe): inferiormost part of the neurohypophysis

  o **INFUNDIBULAR STALK**

  o **MEDIAN EMINENCE:** cone-shaped portion of the hypothalamus

- The neurohypophysis **does NOT make hormones**; it <u>only stores and releases hormones</u> produced in the hypothalamus.

■ The pituitary gland receives its **rich blood supply** from the superior and inferior HYPOPHYSEAL ARTERIES.

  ○ The **superior hypophyseal artery** supplies the entire adenohypophysis and the infundibulum.

  ○ The **inferior hypophyseal artery** supplies the pars nervosa.

■ **The hypothalamus of the brain controls the secretion of hormones from the adenohypophysis in the following manner:**

  ○ First, certain hypothalamic neurons make releasing hormones and inhibiting hormones, which they secrete into a <u>primary capillary plexus</u> in the **median eminence.**

  ○ These hormones then travel through <u>hypophyseal portal veins</u> to a <u>secondary capillary plexus</u> in the **pars distalis.**

  ○ They leave this plexus to <u>signal the adenohypophyseal cells to secrete their hormones</u>, which then enter the secondary capillary plexus and travel to their target cells throughout the body.

■ The NEUROHYPOPHYSIS

  ○ Consists of nervous tissue, which contains the **HYPOTHALAMIC-HYPOPHYSEAL axon TRACT**.

  ○ The cell bodies of the neurons that form this tract are located in the **paraventricular and supraoptic nuclei of the hypothalamus.**

  ○ These neurons synthesize **OXYTOCIN** and **ADH (anti-diuretic hormone, or vasopressin)**, respectively, and store them in their axon terminals in the pars nervosa.

  ○ These stored hormones are then released into capillaries when the neurons fire.

- ▪ The neurohypophyseal hormones have the following functions

  - ○ **ADH** increases resorption of water from the urine and raises blood pressure.

  - ○ **Oxytocin** induces labor and ejection of milk from the breasts.

  - ○ Both these hormones are involved with **social bonding**.

- • THE THYROID GLAND

  - ▪ The thyroid gland, which lies on the superior trachea, consists of spherical follicles covered by epithelial follicle cells and separated by a capillary-rich connective tissue.

    - ○ These follicles are filled with a **colloid of THYROGLOBULIN**, a storage protein containing thyroid hormone.

  - ▪ **Thyroid hormone (TH)**, which **contains iodine and increases basal metabolic rate**, is made continuously by follicle cells.

    - ○ Stored within the follicles until **TSH** from the pituitary gland **signals the follicle cells to reclaim the TH** and secrete it into the extrafollicular capillaries.

  - ▪ **PARAFOLLICULAR CELLS** protrude from the thyroid follicles and **secrete the hormone CALCITONIN**, which can *lower blood calcium concentrations in children.*

- • THE PARATHYROID GLANDS

  - ▪ Several pairs of parathyroid glands lie on the dorsal aspect of the thyroid gland.

  - ▪ Their CHIEF CELLS are arranged in thick, branching cords and secrete **parathyroid hormone (PTH), which raises low blood calcium levels**.

- • THE ADRENAL (SUPRARENAL) GLANDS

  - ▪ The paired adrenal glands lie on the superior surface of each kidney.

    - ○ Each adrenal gland has two distinct parts, an outer CORTEX and an inner MEDULLA.

- The **ADRENAL MEDULLA** consists of spherical clusters (and some branching cords) of **CHROMAFFIN CELLS**.

  o Upon sympathetic stimulation, these cells **secrete epinephrine (E) and norepinephrine (NE)** into the blood – the surge of adrenaline that is experienced during fight-or-flight situations.

- **The ADRENAL CORTEX has three layers**, each of which are descriptively named for its histological structure:

  1. Outer **ZONA GLOMERULOSA**: contains cells arranged in **spherical clusters**

  2. Middle **ZONA FASCICULATA**: cells are arranged in **parallel cords**

  3. Inner **ZONA RETICULARIS**: cells are arranged in a **branching network**

- **The steroid hormones secreted by the adrenal cortex (CORTICOSTEROIDS) include:**

  o **MINERALOCORTICOIDS** (mostly from the *zona glomerulosa*)

    • Mineralocorticoids (**mainly *aldosterone*) conserve water and sodium** by increasing resorption of these substances by the kidney.

  o **GLUCOCORTICOIDS** (mostly from the *zona fasciculata and zona reticularis*)

    • Glucocorticoids (**mainly *cortisol*) help the body cope with stress** by stabilizing blood glucose levels.
    In large quantities, they **also inhibit inflammation and the immune system**.

  o The androgen **DIHYDROEPIANDROSTERONE, DHEA** (from the *zona reticularis*)

    • **The functions of DHEA are unclear but probably beneficial.**

- STEROID-SECRETING CELLS, including the cells in the gonads that secrete sex hormones, have an **abundant SER (smooth ER)**, **tubular *cristae*** in their mitochondria, **abundant lipid droplets**, and **no secretory granules**.

- **THE PINEAL GLAND**

  - The pineal gland, on the roof of the diencephalon, contains **PINEALOCYTES**, which cluster into spherical clumps and cords separated by dense particles of calcium called **PINEAL SAND**.

- **Pinealocytes secrete the hormone MELATONIN, which helps regulate circadian rhythms.**

  o This secretion is signaled by the **suprachiasmatic nucleus of the hypothalamus** through a sympathetic pathway.

- **THE PANCREAS**

  - The ENDOCRINE STRUCTURES in the pancreas are the spherical **PANCREATIC ISLETS**.

    o These islets consist of *alpha (α), beta (ß), delta (D)* and *F (PP) cells*, which are arranged in twisting cords.

  - **ALPHA cells secrete GLUCAGON**, which raises blood sugar levels, whereas **BETA cells secrete INSULIN**, which lowers blood sugar levels.

- **THE THYMUS**

  - The thymus is an important organ of the immune system.

  - It secretes THYMIC HORMONES, which are essential for the production of **T LYMPHOCYTES**.

- **THE GONADS**

  - Various cells in the OVARIES and TESTES **secrete steroid sex hormones, ESTROGENS and ANDROGENS**.

OTHER ENDOCRINE STRUCTURES

- Some **MUSCLE CELLS in the ATRIA** of the HEART secrete **atrial natriuretic peptide (ANP)**, which stimulates loss of body fluids and salts through the production of a sodium-rich urine.

- Endocrine cells are scattered WITHIN the EPITHELIUM of the DIGESTIVE TRACT and OTHER GUT-DERIVED ORGANS (respiratory tract, etc.).

  - These epithelial cells, which have some neuron-like properties, make up the *diffuse neuroendocrine system* (DNES).

  - There are many classes of DNES cells, some of which secrete hormones that regulate digestion.

- The **PLACENTA** secretes hormones of pregnancy.

- The **KIDNEY** secretes RENNIN and ERYTHROPOIETIN.

- The **SKIN** produces vitamin D.

PRACTICE QUESTIONS 4:
The Nervous and Endocrine Systems

1. Which pituitary cell type secretes luteinizing hormone?
   a. corticotropic cells
   b. thyrotropic cells
   c. somatotropic cells
   d. mammotropic cells
   e. gonadotropic cells

2. Respiratory activities are controlled by:
   a. the thalamus.
   b. centers in the pons.
   c. the medulla oblongata.
   d. the hypothalamus.
   e. B and C

3. Conscious thought processes and all intellectual functions originate in the:
   a. cerebellum.
   b. cerebral hemispheres.
   c. limbic system.
   d. temporal lobe.
   e. pons.

4. A highly specialized region of the sympathetic division that causes widespread sympathetic activation is found in the:
   a. adrenal cortex.
   b. liver.
   c. adrenal medulla.
   d. brain.
   e. A and C

5. Typical sympathetic postganglionic fibers that release norepinephrine at neuroeffector junctions are classified as:
   a. adrenergic.
   b. cholinergic.
   c. norepinephric.
   d. nonsecretory.
   e. neuroendocrine.

6. The autonomic division of the nervous system directs:
    a. conscious control of skeletal muscles.   d. emotions.
    b. processes that maintain homeostasis.   e. B and C
    c. behavior.

7. Which gland is comprised of chief cells that are arranged in thick, branching cords and secrete a hormone, which raises low blood calcium levels?
    a. thyroid gland
    b. adrenal gland
    c. parathyroid gland
    d. thymus
    e. pituitary gland

8. The sympathetic division of the ANS differs from the parasympathetic division of the ANS in that:
    a. the sympathetic system has relatively longer preganglionic fibers than does the parasympathetic system.
    b. the sympathetic system promotes rest, relaxation, repose, and nutrient uptake, while the parasympatheitic system increases alertness and metabolism.
    c. the postganglionic fibers of the sympathetic system are relatively longer than those of the parasympathetic system.
    d. the sympathetic system is more divergent than is the parasympathetic system.
    e. C and D

9. Cranial nerves that have ANS fibers are:
    a. VIII, XI, XII.
    b. III, VII, IX, X.
    c. IV, V, VI, VII.
    d. I, II, III.
    e. VII, XI, XII.

10. Which of the following is *not* a branch of the facial nerve?
    a. buccal
    b. temporal
    c. zygomatic
    d. cervical
    e. ophthalmic

11. The ventral root of a spinal nerve contains:
    a. axons of motor neurons.
    b. axons of sensory neurons.
    c. dorsal root ganglia.
    d. cell bodies of motor neurons.
    e. interneurons.

12. Only cones are found in the:
    a. lamina densa.   d. choroid layer.
    b. optic disc.   e. ciliary body.
    c. fovea.

13. The major <u>somatosensory</u> pathways include the:
    a. posterior column pathway.
    b. corticocerebral pathway.
    c. spinocerebellar pathway.
    d. A and C
    e. All of the above.

14. The parasympathetic division of the ANS generally does all of the following *except*:
    a. functions as the "rest and repose" division.
    b. conserves energy.
    c. increases respiratory activities.
    d. promotes sedentary activities.
    e. None of the above. (There are no exceptions.)

15. Which of the following is/(are) a special sensory receptor(s)?
    a. thermoreceptor
    b. auditory receptor
    c. chemoreceptor
    d. mechanoreceptor
    e. A, C, and D

16. The major function of the inferior colliculi includes reflexes involved with:
    a. the taste buds.
    b. the nose.
    c. the eyes.
    d. the ears.
    e. None of the above.

17. To contact an oligodendrocyte, where would a microprobe have to be inserted in the nervous system?
    a. Along any axon in the PNS, myelinated or not
    b. Near or at a neuron cell body in the PNS
    c. In the CNS only
    d. In the adrenal medullae
    e. B and C

18. Cranial nerve IX exits through which foramen?
    a. foramen ovale
    b. foramen spinosum
    c. foramen magnum
    d. jugular foramen.
    e. foramen rotundum.

19. Which cranial nerve(s) exit(s) through the superior orbital fissure?
    a. II, III, IV
    b. III, IV, VI, V_1 (ophthalmic branch)
    c. III, IV, VII, VIII
    d. IV, VI, VII
    e. III, VI, V_2 (maxillary branch)

20. Glucocorticoids, which stabilize blood glucose levels, are secreted by the:
    a. zona glomerulosa.
    b. zona fasciculata.
    c. zona reticulata.
    d. B and C
    e. A and C

21. Which hormones are found in the neurohypophysis?
    a. oxytocin and FSH   d. oxytocin and PRL
    b. LH and ADH   e. FSH and LH
    c. ADH and oxytocin

22. Which disorder(s) reflects damage to auditory receptor cells or neural pathways?
    a. sensorineural deafness
    b. Ménière's syndrome
    c. motion sickness
    d. conduction deafness
    e. A and D

23. Which of the following structures are found in the membranous labyrinth?
    a. vestibule and cochlea
    b. utricle and saccule
    c. semicircular canals
    d. A and C
    e. B and C

24. Which of the following is part of the vascular tunic?
    a. sclera   d. optic nerve
    b. corneal limbus   e. ciliary body
    c. retina

25. Which statement(s) is/are <u>true</u>?
    a. Fast-adapting receptors are phasic.
    b. Slow-adapting receptors are phasic.
    c. Fast-adapting receptors are tonic.
    d. A and B
    e. B and C

26. Which nerve innervates the <u>add</u>uctor muscle groups and superomedial thigh skin?
    a. sciatic nerve
    b. femoral nerve
    c. obturator nerve
    d. common fibular nerve
    e. pudendal nerve

obturator foreman is at the top of the femur/pelvis.

27. Regarding spinal nerves, which structure contains motor fibers that originate in the spinal cord?
    a. dorsal root
    b. ventral root
    c. dorsal ramus
    d. ventral ramus
    e. dermatome

28. Which brachial plexus nerve is formed by the lateral and medial cords?
    a. axillary nerve
    b. ulnar nerve
    c. musculocutaneous nerve
    d. median nerve
    e. radial nerve

29. Which mesencephalic structure(s) acts in the classic "startle response"?
    a. cerebral peduncle
    b. pontine nuclei
    c. red nucleus
    d. substantia nigra
    e. corpora quadrigemina

Chapter 18 – BLOOD

INTRODUCTION

- The **circulatory system** is comprised of the *cardiovascular system* (the heart, blood vessels, and blood) and the *lymphatic system* (lymphatic vessels and lymph).

    - The main function of the cardiovascular system is to transport cells and dissolved materials, including nutrients, wastes, and respiratory gases throughout the body.

- BLOOD is a **specialized <u>fluid</u> connective tissue.**

- ## FUNCTIONS of BLOOD :

    - It distributes nutrients, oxygen (O_2), and hormones to every cell (~75 trillion cells) in the body.

    - It delivers metabolic wastes to the kidneys for excretion.

    - Transports immunological cells that provide defense to peripheral tissues, from pathogens and toxins

    - Stabilizes the pH and electrolyte composition of interstitial fluids

    - Its CLOTTING REACTION restricts the loss of fluid through damaged vessels or tissue injury.

    - Stabilizes the body temperature by absorbing and redistributing heat

COMPOSITION of the BLOOD

- **Two (2) PRIMARY COMPONENTS of blood:**

    - PLASMA – the liquid matrix of blood, which is only slightly denser than water

        - **Contains dissolved proteins**, instead of the collagen, elastic, and reticular fibers found in typical loose connective tissues.

- ▪ **FORMED ELEMENTS** – blood cells and fragments that are suspended in the plasma

  - ○ RED BLOOD CELLS (RBCs), or ERYTHROCYTES

  - ○ WHITE BLOOD CELLS (WBCs), or LEUKOCYTES

  - ○ PLATELETS (sometimes called THROMBOCYTES)

WHOLE BLOOD **:** a mixture of both the plasma and the formed elements

- ▪ For clinical purposes, the components of whole blood may be separated, or **FRACTIONATED**.

- ▪ Whole blood is sticky, cohesive, and resistant to flow, which are characteristics that determine the **VISCOSITY** of a solution.

- ▪ Average of 5-6 liters of whole blood in the cardiovascular system in adult males

- ▪ Average of 4-5 liters of whole blood in adult females

- ▪ Alkaline pH range of 7.35 – 7.45

- ▪ **HYPOVOLEMIC** – low blood volume

- ▪ **NORMOVOLEMIC** – normal blood volume

- ▪ **HYPERVOLEMIC** – excessive (high) blood volume

- • **PLASMA**

  - ▪ Accounts for about 55% of blood volume

  - ▪ Contains ~92% water

  - ▪ **Differences between plasma and interstitial fluid involve the concentrations of dissolved gases and proteins:**

- o *Dissolved oxygen (O_2) concentration in plasma is higher;* dissolved carbon dioxide (CO_2) concentration in interstitial fluid is higher in interstitial fluid

- o *Significant numbers of dissolved proteins in plasma*

  - • Most plasma proteins are **large** and **globular**, preventing them from crossing the capillary walls.

- ▪ **Three CLASSES of PLASMA PROTEINS**

  - o **ALBUMINS** constitute ~60% of the plasma proteins

    - • The *smallest* of the major plasma proteins

    - • Major contributors to the **osmotic pressure** of the plasma

    - • Important role in the transport of *fatty acids*, *steroid hormones*, and other substances

  - o **GLOBULINS** constitute ~35% of the plasma proteins.

    - • **IMMUNOGLOBULINS** (also called *antibodies*) – attack foreign proteins and pathogens

    - • **TRANSPORT GLOBULINS** – bind small ions, hormones, and other compounds, such as insoluble compounds and compounds to be excreted at the kidneys

  - o **FIBRINOGEN** accounts for ~4% of the plasma proteins.

    - • The *largest* of the plasma proteins

    - • Essential for the **clotting reaction**

      - ▪ Fibrinogen molecules interact with each other to form the large, insoluble strands of **fibrin**, under normal conditions.

      - ▪ **FIBRIN** fibers provide the basic framework for a blood clot.

      - ▪ **SERUM** is the fluid that remains if fibrinogen is removed from plasma (*the conversion of fibrinogen to fibrin*).

o When albumins or globulins attach to lipids that are not water-soluble, they form **lipoproteins**.

  • **LIPOPROTEINS** are protein-lipid combinations that readily dissolve in plasma, thus enabling insoluble lipids to be delivered to peripheral tissues.

■ **The primary source of plasma proteins is the liver**, which synthesizes and releases more than 90% of such proteins.

FORMED ELEMENTS

• The major *cellular* components of blood are *erythrocytes* (or *red blood cells*) and *leukocytes* (or *white blood cells*).

• The *non-cellular* formed elements of blood are the *platelets*, which function in the clotting reaction (response).

• ## ERYTHROCYTES, or RED BLOOD CELLS (RBCs)

  ■ Account for slightly less than 50% of the total blood volume

  ■ **HEMATOCRIT** value indicates the percentage of whole blood contributed by formed elements

    o This value closely approximates the volume of RBCs because blood has a content ratio of 1000 RBCs for each WBC.

  ■ **STRUCTURE and CHARACTERISTICS:**

    o **Anucleate, biconcave discs** whose unusual shape provides strength and flexibility

    o The biconcave disc shape also provides a disproportionately large surface area for a cell its size, which permits **rapid diffusion** between the RBC cytoplasm and surrounding plasma.

    o **No** mitochondria or ribosomes present

    o Red color due to presence of hemoglobin molecules

    o Circulation lifespan of approximately 120 days

- Damaged or dead RBCs are recycled by phagocytes.

  o **ROULEAUX** – the stacks (*like stacked dinner plates*) formed by the RBCs due to their biconcave shape, allowing them to pass easily through small vessels

  o **HEMOGLOBIN** molecules account for more than 95% of the proteins of RBCs.

    - Confers RBCs with the ability to transport O_2 and CO_2

    - Structure of hemoglobin: a globular protein formed from **four subunits**

      - Each subunit contains a single molecule of **HEME**, which holds an iron ion that can freely interact (reversible binding) with an oxygen molecule.

- **FUNCTIONS of RBCs:**

  o *Transport dissolved oxygen (O_2) from the lungs to the tissues*

    - Respiratory gas exchange **at the lungs**:

      - CO_2 diffuses out of the blood and O_2 diffuses into the blood.

  o *Transport carbon dioxide (CO_2) from the tissues to the lungs*

    - Respiratory gas exchange **in the peripheral tissues**:

      - O_2 diffuses out of the blood and CO_2 diffuses into the blood.

- **BLOOD TYPES:**

  o Determined by the presence or absence of specific components, called **SURFACE ANTIGENS** (or *agglutinogens*) in the plasmalemmae of RBCs.

  o Three surface antigens of particular importance are assigned as **A**, **B**, and **D (Rh)**.

    - **Type A blood** has surface antigen A.

    - **Type B blood** has surface antigen B.

    - **Type O blood** has neither surface antigen.

- **Rh-positive (Rh+) blood** has the Rh surface antigen (or D surface antigen).

- **Rh-negative (Rh-) blood** does not have the Rh surface antigen.

o **AGGLUTININS** are **antibodies** that are specific to these surface antigens.

o During blood transfusions, **CROSS-REACTION** occurs when antibodies within a person's plasma (the recipient's blood) react with RBCs bearing foreign surface antigens (from the donor's blood).

- Initially, **AGGLUTINATION** occurs in which the RBCs clump together.

- The RBCs may also **HEMOLYZE**, or rupture.

- **COMPATIBILITY** of the blood types of the donor and the recipient avoids this type of cross-reaction.

- ## LEUKOCYTES, or WHITE BLOOD CELLS (WBCs)

  - Leukocytes are scattered throughout the peripheral tissues.

  - WBCs represent only a small fraction of their total population.

  - **FUNCTIONS**:

    o Help defend the body against pathogens

    o Remove toxins, wastes, and abnormal or damaged cells

  - **GENERAL CHARACTERISTICS**:

    o WBCs contain nuclei of characteristic sizes and shapes.

    o They are as large as or larger than erythrocytes (RBCs).

  - ## Two CLASSES of WBCs

    o **GRANULAR LEUKOCYTES** (or **GRANULOCYTES**), which have large granular inclusions in their cytoplasm

o **AGRANULAR LEUKOCYTES** (or **AGRANULOCYTES**), which do not have visible cytoplasmic granules

- **DIFFERENTIAL COUNT** of the WBC population is provided by a stained blood smear.

  o Typical microliter (*μl*) of blood contains 6000-9000 WBCs

  o **LEUKOPENIA**: indicates <u>inadequate</u> numbers of leukocytes; (WBC count of <2500 per *μl* indicates a serious disorder)

  o **LEUKOCYTOSIS**: refers to <u>excessive</u> numbers of leukocytes (WBC count of >30,000 per *μl* indicates a serious disorder)

  o The suffixes (word endings) *–penia* and *–osis* are used to indicate low or high numbers, respectively, of specific types of WBCs.

    <u>For example</u>:
    - *Lympho**penia*** means inadequate numbers of lymphocytes.

    - *Lymphocy**tosis*** means excessive numbers of lymphocytes.

- **DIAPEDESIS**: the ability to move through vessel walls

  o WBCs can migrate across the endothelial lining of a capillary by squeezing between adjacent endothelial cells.

- **CHEMOTAXIS**: the attraction to specific chemical stimuli

  o WBCs are attracted to the chemical signs of inflammation or infection in the adjacent interstitial fluids; thereby, drawing them to invading pathogens, damaged tissues, and other WBCs that are already in the damaged tissues.

- **Three TYPES of GRANULAR LEUKOCYTES (or *granulocytes*)**

  o **NEUTROPHILS** : account for 50-70% of the circulating WBCs

    - Their cytoplasm is packed with **pale, neutral-staining granules** that contain *lysosomal enzymes* and *bactericidal* (antibacterial) compounds; hence, their name.

    - Mature neutrophils have a diameter of 12-15 *μm*, making them nearly twice the size of a RBC.

    - A neutrophil contains a very dense, contorted nucleus.

- **POLYMORPHONUCLEAR LEUKOCYTES** (or **PMNs**) – neutrophils whose nuclei have been condensed into a series of lobes, giving them the appearance of *beads on a string*

- Highly mobile phagocytes, which <u>specialize in attacking and digesting bacteria</u>

- Short life span of about 12 hours

- They are usually the first of the WBCs to arrive at an injury site.

o **EOSINOPHILS (or ACIDOPHILS)** : represent 2-4% of the circulating WBCs

- Their **cytoplasmic granules stain with eosin**, an acidic red dye; hence, their name.

- Similar in size to neutrophils

- Deep red granules in their cytoplasm

- Its **bilobed** (two-lobed) **nucleus** is its hallmark characteristic.

- *Phagocytic cells*, which are attracted to foreign compounds that have reacted with circulating antibodies

- Their presence increase dramatically during an allergic reaction or a parasitic infection.

- They are also attracted to damaged tissues, where they release enzymes that reduce inflammation and control its spread to adjacent tissues.

o **BASOPHILS** : account for <1% of the circulating WBCs

- They have numerous **cytoplasmic granules that stain with basic dyes**; hence, their name.

- Migrate to injury sites and cross the capillary endothelium to accumulate within the damaged tissues, where they **discharge their granules into the interstitial fluids**. Their granules contain:

  - **HISTAMINE** – compound that dilates blood vessels

  - **HEPARIN** – compound that prevents blood from clotting

  - When these granules are released, they **increase the capillary and venule permeability**, resulting in increased inflammation response at the injury site.

236

- Basophils also release chemicals that stimulate mast cells and attract other basophils and other WBCs to the injured area.

- **Two TYPES of AGRANULAR LEUKOCYTES (or *agranulocytes*)**

  o **MONOCYTES** : represent 2-8% of the WBC population

  - The *largest* of the WBCs: 16-20 *μm* in diameter

    - About 2-3x the diameter of a typical RBC

  - Almost *spherical* in shape

  - Relatively easy to identify by their size and the shape of their nucleus, which is typically a **large oval or kidney bean-shaped nucleus**

  - They become **FREE MACROPHAGES** (highly mobile, phagocytic cells) when they leave the bloodstream and enter peripheral tissues.

  - During their phagocytic action, they release chemicals that attract and stimulate other monocytes and other phagocytic cells.

  o **LYMPHOCYTES** : account for 20-30% of the WBC population

  - Typically appears (under light microscopy) as having very little cytoplasm, that forms a thin halo around a relatively large, round purple-staining nucleus

  - Usually slightly larger than RBCs

  - Lymphocytes are the **primary cells of the lymphoid system** (a network of special vessels and organs that are distinct from, but connected to, those of the cardiovascular system).

  - Responsible for **SPECIFIC IMMUNITY** (the body's ability to mount a counterattack against invading pathogens or foreign proteins *on an individual basis*)

  - Three groups of lymphocytes:

    - **T cells** - enter peripheral tissues and attack foreign cells directly

    - **B cells** – differentiates into *plasmocytes* (plasma cells) that secrete antibodies that attack foreign cells or proteins in distant portion of the body

- ■ | NK cells, or ***natural killer cells*** | (or *large granular lymphocytes*) – cells that are responsible for ***immune surveillance***, a process which destroys abnormal tissue cells

   - ■ (T cells and B cells cannot be distinguished with the light microscope.)

Mnemonic Devices: "***Never Let Monsters Eat Babies***" or "***Never Let Monkeys Eat Bananas***" --- to remember the classes of leukocytes in the *decreasing order of their abundance in blood*: **N**eutrophils – **L**ymphocytes – **M**onocytes – **E**osinophils - **B**asophils

---

PLATELETS

- ■ Flattened, **membrane-enclosed packets of cytoplasm**, which appear round when viewed from above and appear spindle-shaped when viewed in section

- ■ They are sometimes referred to as **THROMBOCYTES**; although, platelets technically are <u>not</u> <u>cells</u>.

- ■ **MEGAKARYOCYTES**: enormous cells (in the bone marrow) that shed their cytoplasm in membrane-enclosed packets (platelets), which are released into the blood circulation

   - ○ Up to 160 μm in diameter

   - ○ A mature megakaryocyte produces around 4000 platelets.

- ■ Platelets are **continually replaced**.

- ■ Circulation life span of 10-12 days (removed by phagocytes)

- ■ An average of 350,000 platelets per 1 μl of blood

   - ○ **THROMBOCYTOPENIA**: an **abnormally low** platelet count (80,000 per μl, or less), which indicates excessive platelet destruction or inadequate platelet production

   - ○ **THROMBOCYTOSIS**: an **excessive** platelet count (1,000,000 per μl), which indicates accelerated platelet formation in response to infection, inflammation, or cancer

- **HEMOSTASIS**: a process that prevents the loss of blood through the walls of damaged vessels

  o Platelets are one of the components in a **vascular CLOTTING SYSTEM** that also includes plasma proteins and the cells and tissues of the *circulatory system.*

- **FUNCTIONS of the PLATELETS:**

  o *Transport chemicals that are important to the clotting reaction*

    • Platelets release enzymes and other factors to initiate and control the clotting process

  o *Form a temporary patch in the walls of damaged blood vessels*

    • Platelets form a *platelet plug* by clumping together at the injury site; thereby reducing the rate of blood loss while clotting occurs.

  o *Active contraction after clot formation has occurred in order to reduce the size of the break (pull the cut edges together) in the vessel wall*

    • Platelets contain filaments of *actin* and *myosin* that allow them to produce contractions.

HEMOPOIESIS

- The process of **blood cell formation**

- Embryonic blood cells differentiate into stem cells that produce blood cells by their divisions.

- The bone marrow becomes the primary site of blood cell formation in adults.

- **Stem cells**, called **PLURIPOTENTIAL STEM CELLS (PPSC)**, or *hemocytoblasts*, divide to give rise to all the blood cells.

  - PPSCs give rise to two multipotential stem cell lines:

    o **Multipotential myeloid stem cells** (or **myeloid stem cell**) divide to form five different types of stem cell lines, each with relatively restricted functions.

      • 2 of the stem cell lines produce **RBCs** and **megakaryocytes**.

      • The other 3 stem cell lines give rise to the various forms of **WBCs**.

- ○ **Multipotential lymphoid stem cells** (or **lymphoid stem cell**) divide to form two different types of stem cell lines, each with relatively restricted functions as well.

  - • One stem cell line ultimately forms *plasmocytes* (plasma cells).

  - • The other stem cell line forms **T cells**.

- • ⌈ **ERYTHROPOIESIS** ⌉ : **the formation of erythrocytes, or RBCs**

  - ■ Occurs primarily within the red bone marrow (myeloid tissue) in adults

    - ○ Normal erythropoiesis in myeloid tissues requires adequate supplies of amino acids, iron, and **vitamin B$_{12}$**, a vitamin obtained from dairy products and meat.

  - ■ **ERYTHROPOIESIS-STIMULATING HORMONE** (or **erythropoietin, EPO**): a hormone that regulates erythropoiesis

    - ○ EPO is produced and secreted under **hypoxic** (low-oxygen) conditions, mainly in the kidneys.

    - ○ <u>Two major effects of EPO:</u>

      - • *Stimulates increased rates of cell division in erythroblasts* (immature RBCs) and in the stem cells that produce erythroblasts

      - • *Increases the rate of RBC maturation*, mainly by accelerating the rate of hemoglobin synthesis

  - ■ ⌈ **STAGES in RBC MATURATION** ⌉ :

    - ○ <u>**ERYTHROBLASTS**</u>

      - • Very *immature RBCs* that are actively synthesizing hemoglobin

    - ○ <u>**RETICULOCYTES**</u>

      - • *Mature RBCs* that have just shed their nuclei

      - • This is the last step in the maturation process.

      - • They enter the blood circulation.

      - • Gradually develop the appearance of mature erythrocytes (RBCs)

- **LEUKOPOIESIS** : the production of **leukocytes**, or **WBCs**

  - Stem cells for the production of WBCs originate in the bone marrow.

    - **Granulocytes** complete their development in the bone marrow.

    - **Monocytes** begin their differentiation in the bone marrow.

      - They then enter the blood circulation.

      - Finally, they complete their development when they become **free macrophages** in peripheral tissues.

  - **LYMPHOPOIESIS** : the production of **lymphocytes**

    - Their stem cells originate in the bone marrow also, but many of them subsequently migrate to the **THYMUS**.

    - **PRIMARY LYMPHOID ORGANS** – the bone marrow and thymus – produce daughter cells destined to become specialized lymphocytes.

      - <u>In the bone marrow</u>: produce **immature B cells** and **NK cells**

      - <u>In the thymus</u>: production of **immature T cells**

      - These immature cells may subsequently migrate to **SECONDARY LYMPHOID STRUCTURES**, such as the spleen, tonsils, or lymph nodes.

  - **Regulating factors of lymphocyte maturation** have yet to be fully understood.

    - **Colony-stimulating factors (CSFs)**: collective group of several hormones involved in the regulation of other WBC populations

    - Commercially available CSFs have a clinical use in cancer chemotherapy, in that they stimulate the production of WBCs.

Chapter 19 – THE HEART

OVERVIEW

- The heart is a cone-shaped, **muscular double-pump** with two functions:

    - Its <u>right side receives oxygen-poor blood</u> from body tissues and pumps blood to lungs for oxygenation and dispose of carbon dioxide.

    - Its <u>left side receives the oxygenated blood</u> returning from the lungs and pumps oxygen-rich blood throughout the body to supply oxygen and nutrients.

- *Pulmonary circuit*: blood vessels from the heart that carry blood to and from the lungs

- *Systemic circuit*: blood vessels from the heart that transport blood to and from all body tissues

- **Atria (2)**: receiving chambers of the heart; receive blood from the pulmonary and systemic circuits

- **Ventricles (2)**: main pumping chambers that pump blood around both circuits

STRUCTURE OF THE HEART

- **COVERINGS**: (*Superficial to Deep*)

    Pericardium (encloses the heart): <u>triple-layered</u> sac, which is comprised of *fibrous pericardium and serous pericardium*

    - **FIBROUS PERICARDIUM** (outer layer of sac): dense connective tissue; adheres to diaphragm inferiorly; superiorly fused to roots of the great vessels that enter and exit the heart

    <u>Two layers of **serous pericardium**</u> (double-layered closed sac between the fibrous pericardium and the heart):

    - **PARIETAL LAYER of SEROUS PERICARDIUM** (outer): adheres to inner surface of fibrous pericardium

    - **VISCERAL LAYER of SEROUS PERICARDIUM** (inner) = **EPICARDIUM**: continuous with the parietal layer; lies on the heart and considered a part of the heart wall

- Between the parietal and visceral layers, the pericardial cavity contains serous fluid, which decreases friction created between the beating heart and the outer wall of the pericardial sac.

- | **LAYERS of the HEART WALL** | : *(Superficial to Deep)* – Epicardium – Myocardium - Endocardium

  - **EPICARDIUM** (visceral layer of serous pericardium): often infiltrated with fat, esp. in older people

  - **MYOCARDIUM**: forms the bulk of the heart, consisting mainly of cardiac muscle; the layer that actually contracts; cardiac muscle cells are elongated, circularly and spirally arranged networks of cardiac muscles ("bundles")

  - **ENDOCARDIUM**: sheet of endothelium on a thin layer of connective tissue; lines the heart chambers and makes up the heart valves

- | **HEART CHAMBERS** |

  The right (**R**) and Left (**L**) ATRIA (*superior*) are divided by **INTERATRIAL SEPTUM**.

  The R and L VENTRICLES (*inferior*) are divided by **INTERVENTRICULAR SEPTUM**.

  - | **RIGHT ATRIUM** |: forms the entire right border of the human heart; *receiving chamber* for oxygen-poor blood (from the systemic circuit) via three veins:

    - VESSEL OPENINGS associated with the right atrium:

      - **SUPERIOR VENA CAVA** – from upper body regions, superior to the diaphragm

      - **INFERIOR VENA CAVA** – from lower body regions, inferior to the diaphragm

      - **CORONARY SINUS** – from the heart wall

    - EXTERNAL FEATURES of the R Atrium:

      - **RIGHT AURICLE**: small flap shaped like a dog's ear; projects to the left from the superior corner of the atrium

    - INTERNAL FEATURES of the R Atrium:

      - 2 internal regions of right atrium separated by the **CRISTA TERMINALIS** (C-shaped ridge), which is an important landmark in locating the sites where veins enter the right atrium:

- Smooth-walled posterior region;

- Anterior region lined by horizontal ridges, **PECTINATE MUSCLES**

- **FOSSA OVALIS**: a depression in the interatrial septum that marks the spot where an opening existed in the fetal heart, posterior to the end of the crista terminalis

- **TRICUSPID VALVE** (or *right atrioventricular valve*): the right atrium opens into the right ventricle through this valve, inferiorly and anteriorly

- ### RIGHT VENTRICLE (*pulmonary pump*)

  o Receives blood from the right atrium and pumps it into the pulmonary circuit via pulmonary trunk (arteries)

  o Forms most of the anterior surface of the heart

  o **PULMONARY (semilunar) VALVE**: the right ventricle opens into the pulmonary trunk through this valve, superiorly

  o SPECIAL FEATURES of the INTERIOR of the R Ventricle:

    - **TRABECULAE CARNEAE:** irregular ridges of muscle in the right ventricular walls

    - **PAPILLARY MUSCLES:** cone-shaped projections from the walls

    - **CHORDAE TENDINAE:** strong bands, which project superiorly from the papillary muscles to the flaps (cusps) of the tricuspid valve (*right atrioventricular valve*)

- ### LEFT ATRIUM

  o Makes up most of the *heart's posterior surface*, or **base** of the heart

  o Receives oxygenated blood returning from the lungs through (two pairs) two right and two left PULMONARY VEINS, via the pulmonary circuit

  o **LEFT AURICLE:** the only visible part of the left atrium, anteriorly

  o INTERNAL FEATURE of the L Atrium:

    - The walls are mostly smooth-surfaced with **pectinate muscles** lining the auricle only.

- o   Opens into the left ventricle via the **MITRAL VALVE** (*left atrioventricular valve* or *bicuspid valve*)

- ▪   **LEFT VENTRICLE** (*systemic pump*)

  - o   Dominates the heart's inferior surface; forms the **apex** of the heart

  - o   Pumps oxygenated blood into the systemic circuit

  - o   SPECIAL FEATURES of the INTERIOR of the L ventricle:

    - •   **TRABECULAE CARNEAE**
    - •   **PAPILLARY MUSCLES**
    - •   **CHORDAE TENDINAE**
    - •   Contains cusps of the mitral valve

  - o   Superiorly, the left ventricle opens into the stem artery (the *aorta*) of the systemic circulation via **AORTIC (semilunar) VALVE**

PATHWAY OF BLOOD THROUGH THE HEART

- **OXYGEN-POOR BLOOD enters the heart's RIGHT ATRIUM:**

  - ▪   **From upper body regions** superior to the diaphragm (excluding the heart wall) via the **SUPERIOR VENA CAVA**

  - ▪   **From lower body regions** inferior to the diaphragm via the **INFERIOR VENA CAVA**

  - ▪   **From the heart wall via the CAROTID SINUS.**

- The R ATRIUM contracts and, with the aid of gravity, the oxygen-poor blood is propelled from the right atrium, through the **TRICUSPID VALVE**, to the **RIGHT VENTRICLE**.

- The R VENTRICLE contracts and propels the blood through the **PULMONARY semilunar VALVE** through the **PULMONARY TRUNK** (arteries).

- The oxygen-poor blood flows to the lungs via the **PULMONARY CIRCUIT**, to **become OXYGENATED** and to dispel carbon dioxide.

- The **OXYGEN-RICH (oxygenated) BLOOD** flows back to the heart and enters the **LEFT ATRIUM**, via the **PULMONARY VEINS**.

- The L ATRIUM contracts and propels oxygenated blood through the **MITRAL VALVE** to the **LEFT VENTRICLE**.

- The **LEFT VENTRICLE** contracts and propels blood through the **AORTIC semilunar VALVE** out of the heart through the **AORTA** and its branches.

- This oxygenated blood is pumped throughout the body via the **SYSTEMIC CIRCUIT** to deliver oxygen and nutrients to body tissues.

- After the exchange, the blood becomes oxygen-poor again and returns to the heart's R ATRIUM to continue the cycle through the pulmonary and systemic circuits.

Mnemonic devices: Sequence of blood flow through the valves: *"**T**wo **P**igs **M**auled **A**lex"* or *T, P, M, A* correspond to "**T**ricuspid – **P**ulmonary – **M**itral – **A**ortic Valves"

HEART VALVES

- VALVES ENFORCE THE ONE-WAY FLOW OF BLOOD THROUGH THE HEART in the following order: Atria – Ventricles – Great arteries (pulmonary trunk and aorta)

- ### VALVE STRUCTURE

  - Each heart valve consists of 2 or 3 cusps, flaps of endocardium, reinforced by cores of dense connective tissue

  - **Atrioventricular (AV) valves** are located at junctions of the atria & their respective ventricles:
    - Right atrioventricular valve = **TRICUSPID** valve – 3 cusps
    - Left atrioventricular valve = **MITRAL** valve – 2 cusps

  - **Semilunar valves** are located at the junctions of the ventricles & the great arteries:

- **PULMONARY** semilunar valve (right ventricle to pulmonary trunk) – 3 cusps

- **AORTIC** semilunar valve (left ventricle to aorta) – 3 cusps

- **VALVE FUNCTION**

  - **OPEN: allow blood flow**

  - **SHUT: prevent backflow of blood**

  - AV VALVES prevent the backflow of blood into the atria during ventricular contraction.

    o The chordae tendinae and papillary muscles that attach to the AV valves serve as guy wires by anchoring the cusps in their closed position

    o Prevent backflow into the atria during ventricular contraction

  - SEMILUNAR VALVES prevent backflow from the great arteries into the ventricles.

- **HEART SOUNDS**

  - Closing of the valves causes vibrations in the adjacent blood and heart walls, resulting in *"lub-dup"* sounds:

    o *"lub"* – closing of AV valves at the start of ventricular systole

    o *"dup"* – closing of semilunar valves at the end of ventricular systole

FIBROUS SKELETON OF THE HEART

- Electrically inert, dense connective tissue that lies in the plane between the atria and ventricles

- Surrounds all four heart valves like handcuffs

- **Functions:**

  - Anchors the valve cusps

  - Prevents overdilation of the valve openings as blood flows through

- Point of insertion for cardiac muscle bundles in the atria & ventricles

- Blocks the direct spread of electrical impulses from the atria to the ventricles

THE CARDIAC CYCLE

- **Cardiac cycle:** consists of periods of (atrial and ventricular) systole and (atrial and ventricular) diastole

- **The sequence of contraction *in vivo*: Both the atria always contract together followed by the ventricles simultaneously contracting.**

- **HEARTBEAT**: single sequence of atrial contraction followed by ventricular contraction

- **SYSTOLE (ATRIAL or VENTRICULAR)**: <u>contraction</u> of a heart chamber

- **DIASTOLE (ATRIAL or VENTRICULAR)**: time during which the heart chamber is <u>relaxing and filling</u> with blood

- Both atria and ventricles experience systole and diastole

- The walls of the atria are much thinner than those of ventricles because the atria need to exert little effort, with the aid of gravity, to propel blood inferiorly to the ventricles.

CONDUCTING SYSTEM and INNERVATION

- The intrinsic means by which the heart muscle generates and conducts electrical impulses

- ### CONDUCTING SYSTEM

  - A series of specialized cardiac muscle cells that carry impulses throughout the heart musculature, signaling the heart chambers to contract in the proper sequence

- **<u>Conducting System COMPONENTS:</u>**

  - **SINOATRIAL (SA) NODE:** crescent-shaped mass of muscle cells that lies in the wall of the right atrium, just inferior to the entrance of the superior vena cava

    o The impulse that signals each heartbeat begins at the SA node.

o The SA node sets the basic heart rate by generating 70 – 80 pulses/min = "heart's pacemaker."

- **INTERNODAL FIBERS** (from the SA node to the AV node)

- **ATRIOVENTRICULAR (AV) NODE:** cells are small, but typical cardiac muscle cells; a brief delay of contraction-signaling impulses occurs here to ensure that the ventricles fill completely before contraction

- **AV BUNDLE (Bundle of His):** cells are small, but typical cardiac muscle cells

- **R and L BUNDLE BRANCHES (CRURA)**

- **PURKINJE FIBERS (conduction myofibers):** special, large-diameter, barrel-shaped muscle cells called Purkinje myocytes, containing relatively few myofilaments

  o The large diameter maximizes the speed of impulse conduction.

  o The Purkinje fibers are located between the endocardium and myocardium layers of the ventricle.

- **INNERVATION**

  - Extrinsic neural controls can alter the heart's inherent rate set by the SA node.

  - **VISCERAL SENSORY FIBERS:** nerves supplying the heart

  - **PARASYMPATHETIC FIBERS:** decrease the heart rate; arise as branches of the vagus nerve in the neck and thorax

  - **SYMPATHETIC FIBERS:** increase the rate and force of heart contractions; from the cervical and upper thoracic ganglia

  - All nerves serving the heart pass through the cardiac plexus on the trachea before entering the heart. These autonomic nervous system (ANS) fibers project to the cardiac musculature, but mostly project to the SA and AV nodes and the coronary arteries.

Chapter 20 – BLOOD VESSELS and CIRCULATION

PART 1 : GENERAL CHARACTERISTICS and TYPES OF BLOOD VESSELS

GENERAL CHARACTERISTICS

- <u>Oxygenated</u> blood flows away from the heart through the arteries, which diverge/branch into arterioles, continue into capillaries of the organs.

- <u>Deoxygenated</u> blood leaving capillaries is collected into venules, which merge/converge into larger veins that carry blood towards the heart.

- This pattern of blood vessels applies to both the pulmonary and systemic circulation.

- (<u>Pulmonary</u> <u>circulation</u>: in lungs, oxygen and carbon dioxide, CO_2, are exchanged; <u>Systemic circulation</u>: in the rest of the body, oxygen and nutrients are supplied to the body tissues)

STRUCTURE OF BLOOD VESSEL WALLS (all except the very smallest vessels)

- **<u>Three Tunics</u> or <u>Layers</u>** (*outer/external to inner/lumen*): T. Externa – T. Media – T. Intima

  - **TUNICA EXTERNA:** (outer/external) connective tissue layer consisting of collagen and elastic fibers; cells and fibers run longitudinally

  - **TUNICA MEDIA:** sheets of smooth muscle cells arranged circularly, with circular sheets of elastin and collagen in between

  - **TUNICA INTIMA:** (*inner/lumen*) endothelium = simple squamous epithelium; in vessels that are >1mm diameter, a subendothelial layer (thin layer of connective tissue) lies just external to the endothelium

- **<u>FUNCTIONS of each tunic</u>**

  - **<u>Tunica externa</u>**: protects the vessel, further **_strengthens_** its wall and <u>anchors</u> the vessel to surrounding structures.

- <u>**Tunica media**</u>: *vasoconstriction* (decrease diameter of blood vessel) when smooth muscle cells contract; *vasodilation* (increase diameter) when relaxed; elastin and collagen convey *elasticity* and *strength* for resisting the blood pressure (BP) placed on the vessel wall by each heartbeat

- <u>**Tunica intima**</u>: flat endothelial cells form smooth surface, which *minimizes the friction* of blood moving across them

- ## DIFFERENCES between ARTERIES and VEINS

  - Arteries **conduct blood flow away** from the heart; veins conduct blood towards the heart.

  - **Arterial BP** > Venous BP

  - Veins have **wider lumen** compared to similar-sized arteries.

  - Arteries have **thicker tunica media**.

  - Veins have **thicker tunica externa**.

  - Veins have **valves**; arteries do not have valves.

  - **Walls** of arteries are generally **thicker** (retain their round shape) compared to those of veins (tend to collapse).

  - **Elastic laminae**, sing. = *lamina*, (inner and external) exist in (muscular) arteries only, not in veins.

TYPES OF BLOOD VESSELS

- Three types of blood vessels: Arteries, Capillaries, and Veins

- ## ARTERIES – 3 SUBTYPES:
  (*Elastic* Arteries - *Muscular* Arteries - *Arterioles*)

  - Conduct blood away from the heart

  - <u>In the systemic circuit</u>: arteries carry *oxygen-rich blood* to capillaries of organs/tissues

  - <u>In the pulmonary circuit</u>: arteries carry *oxygen-poor blood* to the lungs

  - The TUNICA MEDIA is THICKER than the tunica externa.

- **ELASTIC ARTERIES** (aorta and its major branches)

  o The largest arteries (with larger lumen) are near the heart.

  o Low-resistance conduits = conducting arteries

  o Increased elastin in walls; thick elastin sheets in tunica media

  o High elastin content dampens the surges of BP

- **MUSCULAR ARTERIES** (*distributing arteries*)

  o Lie distal to the elastic arteries

  o Supply organs

  o "Muscular" because tunica media is thicker

  o **UNIQUE FEATURE of muscular arteries:** especially thick sheets of elastin lie on each side of the tunica media

    • **INTERNAL ELASTIC LAMINA** between the tunica intima and tunica media

    • **EXTERNAL ELASTIC LAMINA** between the tunica media and tunica externa

- **ARTERIOLES**

  o Smallest of the arteries

  o Low-resistance conduits = conducting arteries (0.3 mm – 10 μm in diameter)

  o Tunica media consists of only 1-2 layers of smooth muscle cells

  o Larger arterioles: 3 tunics + internal elastic lamina

  o Smaller arterioles: comprised of just a layer of smooth muscle cells + underlying endothelium

  o Regulate amount of blood flow to capillary bed and regulate systemic BP via changing arteriole diameter

- **CAPILLARIES** – 3 TYPES:

  (*Continuous - Fenestrated - Sinusoidal* Capillaries)

  - Smallest of the blood vessels (8-10 μm) just large enough for erythrocytes (red blood cells, RBCs).

  - STRUCTURE: a single layer of endothelium is surrounded by basal lamina

  - MOST IMPORTANT blood vessel type because they renew and refresh interstitial fluid

  - FUNCTIONS of CAPILLARIES:

    o Oxygen and nutrients delivered

    o CO_2 and nitrogenous wastes removed

    o In the lungs, oxygen enters the blood and CO_2 leaves it.

    o In the small intestine, capillaries receive digested nutrients.

    o In endocrine glands, capillaries pick up hormones.

    o In kidneys, nitrogenous wastes are removed from the blood.

  - **CAPILLARY BEDS** : structures by which capillaries supply cells in tissues; networks of the body's smallest blood vessels

    o **METARTERIOLE:** structural intermediate *between arteriole and capillary*

    o **THOROUGHFARE CHANNEL:** structural intermediate *between capillary and venule*; proceeds to join venules, receive true capillaries along the way

    o **PRECAPILLARY SPHINCTERS:** *smooth muscle cells* that wrap around the root of each true capillary where it leaves the metarteriole

      • Function to regulate blood flow to tissue, by opening and closing, according to the tissue's needs

    o Most tissues and organs contain a rich capillary supply.

○ Tendons and ligaments - poor capillary supply

○ Epithelia and cartilage – no capillary supply; blood supply from connective tissues

○ Cornea and the lens – no capillary supply; nourished by aqueous humor and other sources

- **CAPILLARY PERMEABILITY**

○ ANATOMICAL BASIS of how substances are delivered and picked up

○ Capillary structure is well suited for their **function in the exchange of nutrients and wastes** between the blood and the tissues through the tissue fluid.

○ **TIGHT JUNCTIONS**: block passage of small molecules

○ **INTERCELLULAR CLEFTS**: small molecules exit and enter the capillary

○ **PERICYTES**: spider-shaped cells whose thin processes form a widely-spaced network that does not interfere with capillary permeability

  • **Pericytes strengthen and stabilize the capillaries.**

- **THREE TYPES of CAPILLARIES**

○ **CONTINUOUS CAPILLARIES**: more common type; endothelial cells are connected by tight junctions; occur in most organs such as skeletal muscles, skin and the central nervous system (CNS)

○ **FENESTRATED CAPILLARIES**: FENESTRATIONS (PORES) span the endothelium; occur only where there are exceptionally high rates of exchange (e.g. small intestine and synovial membrane of joints where many water molecules exit the blood to form synovial fluid)

○ **SINUSOIDS** (*sinusoidal capillaries*)

  • Wide, leaky, twisted capillaries containing both expanded and narrowed regions

- Usually fenestrated

- Endothelial cells with fewer cell junctions

- Intercellular clefts wide open in some

- Occur wherever extensive exchange of large materials, proteins or cells occur (e.g. bone marrow and spleen)

- <u>FUNCTION</u>: decrease blood flow rate to allow time for the many exchanges that occur across organ walls

- **FOUR ROUTES of CAPILLARY PERMEABILITY**

  o DIRECT DIFFUSION (CO_2 and O_2) across the endothelium

  o INTERCELLULAR CLEFTS allow most of small molecules to enter and exit capillaries

  o CAVEOLAE, singular = *caveola* (*cytoplasmic vesicles*) allow larger molecules to enter and exit

  o FENESTRATIONS in fenestrated capillaries allow some small molecules to enter and exit

- **LOW-PERMEABILITY Capillaries exist in the BLOOD BRAIN BARRIER (BBB)**

  o Prevents all but the most vital molecules from leaving the blood and entering brain tissue

  o Lack of structural features that account for capillary permeability

  o **COMPLETE TIGHT JUNCTIONS** in brain capillaries

  o **NO INTERCELLULAR CLEFTS**

  o **CONTINUOUS capillaries**, not fenestrated

  o **LACK CAVEOLAE** (*cytoplasmic vesicles*); singular = ***caveola***

  o Uncharged and lipid-soluble molecules (O_2, CO_2 and some anesthetics) may cross the BBB.

- **VEINS** – 3 TYPES:

  (*Venules / Veins / Vena Cavae*)

  - Conduct blood from capillaries toward the heart

  - In *systemic circuit*: O_2 –poor blood from capillaries

  - In *pulmonary circuit*: O_2 -rich blood from the lungs

  - Venous BP < Arterial BP

  - **VENULES:**

    o Smallest of the veins (8-100 μm)

    o 1-2 layers of smooth muscle cells in tunica media + thin tunica externa

    o **POSTCAPILLARY VENULES** are the smallest venules, which are comprised of endothelium on which lie pericytes.

    o They function like capillaries.

  - Venules join or merge to form **VEINS**:

    o Veins hold fully 65% of body's blood.

    o TUNICA EXTERNA is THICKER than tunica media

  - **VENA CAVAE (singular = *vena cava*):** largest of the veins

    o Its TUNICA EXTERNA is FURTHER THICKENED by longitudinal bands of smooth muscle.

    o DECREASED ELASTIN content compared to arteries

  - **Several MECHANISMS COUNTERACT the LOW VENOUS BP** and HELP BLOOD FLOW along its course towards the heart:

    o **VALVES:** prevent backflow of blood away from the heart

      - Valves are **cusps formed from *tunica intima*.**

- Most abundant in veins of the limbs because superior direction of venous blood flow is opposed by gravity

- Valves found in veins of the head and neck

- None found in thoracic and abdominal cavities

o **NORMAL MOVEMENT** of our body and limbs

o **SKELETAL MUSCULAR PUMP:** contracting muscles press against veins, forcing valves proximal to area of contraction open and propelling blood towards the heart

o **VENOUS BP is LOWER DUE TO:**
  - Thinner walls
  - Wider lumen
  - Thinner tunica media
  - Thicker tunica externa

VASCULAR ANASTOMOSES

(singular = *anastomosis*)

- Collateral channels where vessels unite or interconnect

- PROVIDE ALTERNATE PATHWAYS for blood to reach a given body region

- **ARTERIAL anastomoses:**

  - Most organs receive blood from >1 arterial branch, and neighboring arteries often communicate with one another to form anastomoses

  - Arterial anastomoses occur around joints where active body movements may hinder blood flow through one channel (e.g. abdominal organs, brain, heart)

  - Occlusion causes severe tissue damage in some organs

- **VENOUS anastomoses:**

  - They occur more freely than arterial anastomoses.

  - Visibly located through the skin on dorsum of hand

  - Occlusion rarely blocks blood flow or leads to tissue death

VASA VASORUM

- **LITTLE VESSELS** (arteries, capillaries and veins) **in the TUNICA EXTERNA of LARGER ARTERIES and VEINS**, which function as the BLOOD SUPPLY of BLOOD VESSEL WALLS

- **FUNCTION: Nourish the outer half of the wall of the larger vessel** (the inner half is nourished by blood in the vessel's lumen)

- No vasa vasorum found in small blood vessels

- May arise as tiny branches from the same large vessel or as small branches from other, nearby, arteries and veins

PART 2 : CIRCULATION of the BODY

THE PULMONARY CIRCULATION

- Begins as O_2-poor blood leaves the right ventricle of the heart via the **PULMONARY TRUNK**, which exits anterior to the aorta and SPLITS INTO **RIGHT and LEFT PULMONARY ARTERIES**, inferior to the aortic arch.

- <u>Each pulmonary artery enters each lung, at the medial aspect, and branches in the lungs as follows:</u>

  - **LOBAR ARTERIES** – repeated branching of arteries decrease in size and branch along with respiratory bronchi –
  - **ARTERIOLES** –
  - **PULMONARY CAPILLARIES** that surround the alveoli, where gas exchange occurs across the respiratory membrane –
  - arising from the **CAPILLARY BEDS**,
  - **PULMONARY VENOUS TRIBUTARIES (*venules*)** empty newly oxygenated blood into –
  - **SUPERIOR and INFERIOR PULMONARY VEINS**, which exit the medial aspect of each lung –
  - each pair of **PULMONARY VEINS** EXTEND FROM THE LUNGS TO THE LEFT ATRIUM of the HEART.

THE SYSTEMIC CIRCULATION

- The circuit begins with O_2-rich blood leaving the left ventricle of the heart through the **AORTA (ascending aorta, aortic arch, descending aorta)**, which progressively decrease in size in the following order: **ELASTIC ARTERIES, MUSCULAR ARTERIES and ARTERIOLES**.

- Arterioles become **METARTERIOLES** in capillary beds, which nourish body tissues with oxygen and nutrients and thus blood becomes O_2-poor.

- Blood flows through **THOROUGHFARE CHANNELS**, proceeds to join **veins** and ends with the two large vena cavae (**SUPERIOR and INFERIOR VENA CAVAE**) and the **CORONARY SINUS**, which empty into the right atrium of the heart.

- | **The SYSTEMIC ARTERIES leave the heart and begin as the AORTA** |, which is divided into three parts: (***Ascending Aorta / Aortic Arch / Descending Aorta***)

    - | **ASCENDING AORTA** |

        o Branches into 2 CORONARY ARTERIES that supply the heart wall

    - | **AORTIC ARCH** |

        o Branches into 3 ARTERIES that run superiorly and supply the head, neck, upper limbs and superior part of the thoracic wall: (*Brachiocephalic Trunk – Left Common Carotid Artery – Left Subclavian Artery*)

        a. **BRACHIOCEPHALIC TRUNK** ascends to the right toward the base of the neck and <u>further divides</u> into:
            - the RIGHT COMMON CAROTID ARTERY and
            - the RIGHT SUBCLAVIAN ARTERY

        b. **LEFT COMMON CAROTID ARTERY**

        c. **LEFT SUBCLAVIAN ARTERY**

        o The R and L SUBCLAVIAN ARTERIES give off branches (**VERTEBRAL ARTERY, THYROCERVICAL TRUNK** and **COSTOCERVICAL TRUNK**) to the neck and run laterally onto the first rib, underlying the clavicle and continue laterally to supply the upper limbs.

        o The sections of the subclavian arteries **exit the thoracic cavity, passing over the outer border of the first rib**, and continue through the upper limb and are <u>named according to the specific region of the upper limb it supplies</u>: AXILLARY ARTERIES – BRACHIAL ARTERIES – RADIAL ARTERIES – ULNAR ARTERIES – PALMAR ARCHES – DIGITAL ARTERIES

        o Most parts of the head and neck are supplied by the **COMMON CAROTID ARTERIES (R and L)**, which divide into an **EXTERNAL** and **INTERNAL** carotid artery EACH.

o Both the R and L **EXTERNAL CAROTID** ARTERIES supply most tissues of the head external to the brain and orbit and send branches to:

- The thyroid gland and larynx *(superior thyroid artery)*

- The tongue *(lingual artery)*

- The skin and muscles of the anterior face *(facial artery)*

- The posterior part of the scalp *(occipital artery)*.

- Each external carotid ends near the temporomandibular joint by splitting into:

  ▪ The *superficial temporal artery* (supplies most of the scalp)

  ▪ The *maxillary artery* (branches also to the teeth, cheeks, nasal cavity and muscles of mastication)

o The **INTERNAL CAROTID** ARTERIES supply the orbits and most of the cerebrum, enters the sella turcica of sphenoid bone just posterior to the foramen and branches off as:

- The *ophthalmic artery* (supplies eyes and orbits)

- Divides into the *anterior and middle cerebral arteries* (supply over 80% of the cerebrum)

- Each anterior cerebral artery anastomoses (unites or interconnects) with each other through a short *anterior communicating artery*.

o The posterior brain is supplied by the **R and L VERTEBRAL ARTERIES**, which ascend the vertebral column, sending branches to the vertebrae and cervical spinal cord and join to form the **unpaired BASILAR ARTERY** (branches to the cerebellum, pons and inner ear) within the cranium.

- At the border of the pons and midbrain, it divides into a pair of *posterior cerebral arteries* (supply the occipital lobes and inferomedial parts of temporal lobes) and is connected to the middle cerebral arteries anteriorly by the *posterior communicating arteries*, which, together with the *anterior communicating artery*, complete the formation of an *arterial anastomosis* called the **cerebral arterial circle**.

- The **CEREBRAL ARTERIAL CIRCLE** (or **CIRCLE of WILLIS**):

  ▪ Forms a loop around the pituitary gland and optic chiasma

- Unites the brain's anterior and posterior blood supplies provided by the internal carotid and vertebral arteries

- This anastomosis **provides alternate routes for blood** to reach brain areas that are affected if either a carotid or vertebral artery becomes occluded.

- ## DESCENDING AORTA

    o Runs posterior to the heart and inferiorly on the bodies of the thoracic and lumbar vertebrae

The descending aorta has two sections:

o **THORACIC AORTA (T_5-T_{12})** supplies the thoracic organs and body wall

o **ABDOMINAL AORTA (T_{12}-L_4)** supplies the abdominal organs and divides into the **R and L COMMON ILIAC ARTERIES**, which supply the pelvis and lower limbs.

- The **common iliac arteries descend each limb** and each section is <u>named according to the specific region of the lower limb it supplies</u>:

    - FEMORAL ARTERIES – POPLITEAL ARTERIES – ANTERIOR and POSTERIOR TIBIAL ARTERIES, et. al.

- **Branches to the body wall, the kidneys,** and other structures outside the peritoneal cavity are **<u>paired</u>**.

    - They originate along the <u>lateral</u> surfaces of the abdominal aorta.

- The **major branches (of the common iliac arteries) to visceral organs are <u>unpaired</u>**, and arise on the <u>anterior</u> surface of the abdominal aorta and they extend into the mesenteries to reach the visceral organs.

    The three unpaired arteries (*superior* to *inferior*) are the celiac trunk, superior and inferior mesentery arteries:

    - The **CELIAC TRUNK** – superiorly located branch that delivers blood to the liver, stomach, esophagus, gallbladder, duodenum, pancreas, and spleen; divides into **three branches:**

        o LEFT GASTRIC ARTERY – supplies stomach and inferior portion of esophagus

        o SPLENIC ARTERY – supplies the spleen and arteries to the stomach and pancreas o COMMON HEPATIC ARTERY – supplies arteries to the liver, stomach, gallbladder, and duodenal area

  ■ The **SUPERIOR MESENTERIC ARTERY** – arises about 2.5 cm inferior to the celiac trunk to supply arteries to the pancreas and duodenum, small intestine, and most of the large intestine

  ■ The **INFERIOR MESENTERIC ARTERY** – delivers blood to the terminal portions of the colon and the rectum

- | The SYSTEMIC VEINS mostly run with corresponding arteries |, but there are important differences in the distributions of arteries and veins.

  ■ The **SUPERIOR VENA CAVA (SVC)** arises from the merging of the following:

   o The **R** (vertical) and **L** (nearly horizontal and longer) **BRACHIOCEPHALIC VEINS** (posterior to the manubrium) are each formed by the union of an **INTERNAL JUGULAR VEIN** and a **SUBCLAVIAN VEIN**

  ■ The **INFERIOR VENA CAVA (IVC)** begins inferiorly as the union of the two COMMON ILIAC VEINS at the level of vertebra L_5.

   o Blood returning from the abdominopelvic viscera and the abdominal wall reaches the heart via the IVC.

   o Most venous tributaries of the IVC share the names of the corresponding arteries.

  ■ **Veins of the HEAD and NECK**: Most blood draining from the head and neck enters 3 pairs of veins.

   o INTERNAL JUGULAR VEINS from the DURAL SINUSES
   o EXTERNAL JUGULAR VEINS
   o VERTEBRAL VEINS

  ■ **Veins of the UPPER LIMBS** are either deep or superficial:

DEEP Veins
   o PALMAR VENOUS ARCHES (DEEP and SUPERFICIAL) supply the hand and empty into
   o the RADIAL and ULNAR VEINS of the forearm, which unite to form
   o the BRACHIAL VEINS of the arm and enter the axilla as

- o the AXILLARY VEINS, which become
- o the SUBCLAVIAN VEIN at the first rib.

SUPERFICIAL Veins
- o CEPHALIC VEINS
- o BASILIC VEINS
- o MEDIAN CUBITAL VEINS
- o MEDIAN VEINS of the FOREARM

- **Veins of the THORAX** empty blood from the first few intercostals spaces into the BRACHIOCEPHALIC VEINS, whereas blood from the other intercostals spaces and from some of the thoracic viscera drains into the **azygos system**:

  - o The AZYGOS SYSTEM of VEINS is a group of veins, which flank the vertebral column and ultimately empty into the SVC.

  - o This system consists of the AZYGOS VEIN, HEMIAZYGOS VEIN and the ACCESSORY HEMIAZYGOS VEIN.

- **Veins of the ABDOMEN**:
  - o LUMBAR VEINS
  - o GONADAL VEINS
  - o RENAL VEINS
  - o SUPRARENAL VEINS
  - o HEPATIC VEINS

  - o **Portal system:** set of vessels in which two capillary beds, interconnected by a vein, lie between the initial artery and the final vein

    - **HEPATIC PORTAL SYSTEM:** special subcirculation that drains the digestive organs of the abdomen and pelvis

- **Veins of the PELVIS and LOWER LIMBS** are either deep or superficial. They ascend and merge to ultimately become the IVC:

DEEP VEINS of the pelvis and lower limbs share the names of the arteries they accompany:
- o PLANTAR VEINS
- o TIBIAL VEINS (ANTERIOR and POSTERIOR)
- o FIBULAR VEINS
- o POPLITEAL VEINS
- o FEMORAL VEINS
- o INTERNAL and EXTERNAL ILIAC VEINS
- o COMMON ILIAC VEINS

SUPERFICIAL VEINS of the pelvis and lower limbs:

o   DORSAL VENOUS ARCH of the foot
o   GREAT SAPHENOUS VEINS empty into the femoral vein
o   SMALL SAPHENOUS VEINS empty into the popliteal vein

Chapter 21 – THE LYMPHOID (LYMPHATIC and IMMUNE) SYSTEM

INTRODUCTION

- The lymphoid system plays a central role in the body's defenses against viruses, bacteria and other microorganisms.

    - The <u>lymphatic system</u> consists of lymphatic circulatory vessels that carry lymph.

    - The <u>immune system</u> contains the lymphocytes, lymphoid tissue, and lymphoid organs, which are involved in the body's fight against disease.

Lymphoid structures can be classified as ***primary*** (*containing stem cells*) or ***secondary*** (*containing immature or activated lymphocytes*).

THE LYMPHATIC SYSTEM

- **LYMPH** is *excess tissue fluid*, which is similar to plasma but contains a *lower concentration of proteins.*

- **LYMPHATICS**, or ***lymphatic vessels*** pick up lymph and return it to the great veins at the root of the neck.

    - Lymphatics <u>interconnect</u> the lymphoid organs and lymphoid tissues.

    - Lymphatics produce, maintain and distribute **lymphocytes** (cells that attack invading organisms, abnormal cells, and foreign proteins).

    - Lymphatic vessels <u>help maintain blood volume</u> and eliminate local variations in the composition of the interstitial fluid.

    - Lymphatic vessels also <u>retrieve blood proteins that leak from capillaries</u> and return these proteins to the bloodstream.

    - The vessels of the lymphatic system, from smallest to largest, are **lymphatic capillaries, lymphatic collecting vessels** (with lymph nodes), **lymph trunks,** and **lymph ducts.**

- **LYMPHATIC CAPILLARIES** : (*terminal lymphatics*)

  - Lymphatic capillaries weave through the loose connective tissues of the body.

    - These closed-end tubes are **highly permeable** to entering tissue fluid and proteins because their endothelial cells are loosely joined and overlap to act as a one-way valve, preventing fluid from returning to the intercellular spaces.

    - Disease-causing microorganisms and cancer cells also enter the permeable lymphatic capillaries and spread widely through the lymph vessels.

  - Lymphatic capillaries called **LACTEALS** absorb digested fat from the small intestine.

- **MAJOR LYMPHATIC COLLECTING VESSELS**

  - **Superficial** and **deep** lymphatic collecting vessels run alongside arteries and veins.

    - The difference is that they have **thinner walls** and many **more valves** than do veins.

    - As a result of the closely spaced valves within these collecting vessels, they resemble a "*string of beads.*"

  - **Lymph flows very slowly through lymphatic collecting vessels and empty into the THORACIC DUCT and the RIGHT LYMPHATIC DUCT.**

    - Normal body movements, contractions of skeletal muscles, arterial pulsations, and contraction of smooth muscle in the wall of the lymphatic vessel maintain flow.

    - **Lymphatic VALVES prevent backflow.**

- **LYMPH NODES**

  - Clustered along the lymphatic collecting vessels, **bean-shaped lymph nodes remove infectious agents and cancer cells from the lymph stream.**

    - Lymph enters the node via **AFFERENT lymphatic vessels** and exits via **EFFERENT vessels** at the **HILUM**.

    - In between, the lymph percolates through lymph sinuses, where macrophages remove lymph-borne pathogens.

- ### LYMPH TRUNKS

  - The lymph trunks (**lumbar, intestinal, bronchomediastinal, subclavian, and jugular**) each drain a large body region.

  - **All except the intestinal trunk are PAIRED.**

- ### LYMPH DUCTS

  - The **RIGHT LYMPHATIC DUCT** (and/or the nearby trunks) drains lymph from the superior right quarter of the body.

  - The **THORACIC DUCT** (and/or the nearby trunks) drains lymph from the rest of the body.

    o These two ducts empty into the junction of the internal jugular and subclavian veins.

    o The thoracic duct starts at the **cisterna chyli at L_1-L_2** and ascends along the thoracic vertebral bodies.

THE IMMUNE SYSTEM

- Lymphoid organs and lymphoid tissues house millions of **lymphocytes,** important cells of the immune system that recognize specific antigens.

- ### LYMPHOCYTES (3 Types: B cells, T cells, and NK cells)

  - **B and T lymphocytes fight infectious microorganisms in the loose and lymphoid connective tissues of the body.**

    o ### B cells (*bone marrow-derived*) produce antibody-secreting plasma cells.

      - B cells and ANTIBODIES are best at destroying bacteria and bacterial products.

      - B cells can differentiate into PLASMOCYTES, which produce and secrete antibodies that react with specific chemical targets called ANTIGENS.

      - IMMUNOGLOBULINS are antibodies in body fluids.

      - B cells are responsible for **antibody-mediated immunity**.

      - **MEMORY B cells** are activated if the antigen appears again at a later date.

o ⬚ **CYTOTOXIC (CD8⁺) T cells** *(thymus-dependent)* directly kill antigen-bearing cells, which are foreign cells or body cells that have been infected by viruses.

- T cells are best at destroying eukaryotic cells that express surface antigens, such as virus-infected cells and grafted and tumor cells.

- T cells provide **cell-mediated immunity**.

- **REGULATORY T cells** (*helper* and *suppressor*) regulate and coordinate the immune response.

- **MEMORY T cells** are activated if the antigen appears again at a later date.

▪ ⬚ **NATURAL KILLER (NK) CELLS** , (or *large granular lymphocytes*) do not recognize specific antigens but rapidly attack and kill foreign cells, tumor (cancer) cells and virus-infected cells.

o NK cells provide **immunological surveillance**.

▪ Mature lymphocytes patrol connective tissues throughout the body by passing in and out of the circulatory vessels (**recirculation**).

▪ Lymphocytes have a long life span.

• ⬚ **LYMPHOCYTE ACTIVATION and THE IMMUNE RESPONSE**

▪ **Lymphocytes arise from stem cells in the bone marrow.**

o **T cells** develop immunocompetence in the **thymus**.

o **B cells** develop immunocompetence in the **bone marrow**.

o **IMMUNOCOMPETENCE**: the ability to recognize antigens.

- Millions of different lymphocytes, which retain the ability to divide, allow the body to be prepared for any antigen.

o Immunocompetent lymphocytes then circulate to the loose and lymphoid connective tissues, where antigen binding (the antigen challenge) leads to **lymphocyte ACTIVATION**.

- **GOAL of the immune response:** destruction or inactivation of pathogens, abnormal cells, and foreign molecules such as toxins.

- The **ANTIGEN CHALLENGE** involves an interaction among the following three factors:

  (1) The **lymphocyte being activated**

  (2) An **antigen-presenting cell** (dendritic cell or macrophage), and

  (3) A **HELPER (CD4⁺) T lymphocyte**

  ### EVENTS in the ANTIGEN CHALLENGE:

  - Antigen-presenting cells result when **antigens are engulfed by macrophages**, which then **present the antigen(s) to T cells** so they can begin **differentiating**.

  - A newly activated T cell or B cell divides quickly to produce **many short-lived effector lymphocytes** and **some long-lived memory lymphocytes**.

  - Recirculating memory lymphocytes provide **long-term immunity**.

- ### LYMPHOID TISSUE

  - **LYMPHOPOIESIS** (lymphocyte production) involves the bone marrow, thymus, and peripheral lymphoid tissues.

  - **Lymphoid tissue is an often-infected reticular connective tissue in which many B and T lymphocytes gather to fight pathogens or become activated.**

    - It is located in the <u>mucous membranes</u> (as **MALT – mucosa associated lymphoid tissue**) and in the <u>lymphoid organs</u> (<u>except the thymus</u>).

  - Lymphoid tissue **contains lymphoid follicles (lymphoid nodules) with germinal centers**, in which lymphocytes are densely packed in an area of loose connective tissue.

    - Each follicle contains thousands of **B-lymphocytes**, all derived from one activated B cell.

- ☐ **LYMPHOID ORGANS** (Thymus – Lymph Node – Spleen – Tonsils or Aggregated Lymph Nodules)

  - The ☐ **THYMUS** , located in the superoanterior thorax and neck (in the superior mediastinum and posterior to the manubrium), is a primary lymphoid organ that is <u>most active during youth</u>.

    - Its **thymic hormones**, which are secreted by epithelial (reticular) cells, **signal the contained T lymphocytes to differentiate and gain immunocompetence.**

    - **BLOOD-THYMUS BARRIER** – does not allow free exchange between the interstitial fluid and the circulation; thus, protecting the T cells from being prematurely activated

    - The thymus has **lobules**, each with an **outer cortex** packed with <u>maturing T cells</u> and **inner medulla** containing <u>fewer T cells</u> and <u>degenerative thymic (Hassall's) corpuscles</u>.

    - The thymus undergoes **INVOLUTION** – decreases in size (gradually) after puberty.

    - The thymus <u>neither directly fights antigens nor contains true lymphoid tissue</u>.

  - Within a ☐ **LYMPH NODE** , encapsulated masses of lymphoid tissue lie between the sinuses.

    - **DEEP CORTEX** – dominated by **T cells**

    - **OUTER CORTEX and MEDULLA** – contain **B cells** arranged into medullary cords

    - **Lymph glands** – the largest lymph nodes, found where peripheral lymphatics connect with the trunk.
      - **LOCALES:** cervical, axillary, popliteal, inguinal, thoracic abdominal, intestinal and mesenterial lymph nodes serve to protect the vulnerable areas of the body

    - This lymphoid tissue **receives some of the antigens that pass through the node**, leading to lymphocytes activation and memory-lymphocytes production.

  - The ☐ **SPLEEN** has two main functions:

    - The adult spleen contains the largest mass of lymphoid tissue in the body.

    - **Diaphragmatic surface** of the spleen lies against the diaphragm

    - **Visceral surface** lies against the stomach and kidney, and contains a groove called the **HILUM** o   The cellular components form the **PULP** of the spleen.

o   The spleen **removes antigens from the blood**, which is performed by the <u>white pulp</u>.

   •   **WHITE PULP** of the spleen consists of sleeves of lymphoid tissue, each surrounding a central artery.

   •   White pulp resembles lymphoid nodules.

   •   It is surrounded by a high concentration of macrophages.

o   The spleen **destroys worn-out red blood cells (RBCs)**, which is performed by the <u>red pulp</u>.

   •   **RED PULP** of the spleen consists of venous sinuses and strips of blood-filled reticular connective tissue called **splenic cords**, whose macrophages remove worn-out blood cells.

   •   Red pulp contains large number of RBCs with lymphocytes scattered throughout.

▪   The TONSILS in the pharynx, **AGGREGATED LYMPHOID NODULES** in the small intestine, and the **WALL of the APPENDIX** <u>are parts of MALT</u> in which the lymphoid tissue contains an exceptionally high concentration of lymphocytes and follicles.

PRACTICE QUESTIONS 5:
The Cardiovascular and Lymphoid Systems

1. The right ventricle:
   a. receives blood from the right atrium through the *mitral* (*bicuspid*) valve.
   b. has thicker muscular walls than does the left ventricle.
   c. pumps blood out of the heart to the pulmonary circuit.
   d. sends blood out through the aortic semilunar valve to the systemic circulation.
   e. receives oxygenated blood.

2. The superior surface of the heart is the:
   a. sternocostal surface.
   b. anterior surface.
   c. diaphragmatic surface.
   d. base.
   e. apex.

3. The prominent muscles that run along the inner surface of the ventricles and which are attached to *chordae tendinae* are:
   a. pectinate muscles.
   b. chordae tendineae.
   c. fossa ovalis.
   d. trabeculae carneae.
   e. papillary muscles.

4. Besides transporting respiratory gases, what other functions are performed by erythrocytes?
   a. They assist in phagocytizing foreign pathogens.
   b. None; transporting oxygen and carbon dioxide is their only function.
   c. They participate in the immune response with white blood cells.
   d. They carry antigens to the peripheral tissues.
   e. A and B

5. Neutrophils and monocytes are types of:
   a. leukocytes.
   b. basophils.
   c. eosinophils.
   d. erythrocytes.
   e. lymphocytes.

6. The bulges that give the lymphatic vessels the *"string of beads"* appearance is due to:
   a. thickenings in the endothelial lining of the lymph vessels at evenly spaced intervals.

    b.   periodic inflammation of the tissues that remove bacteria and viruses form other body tissues.

    c.   regions of higher pressure within the lymph vessels, which cause the walls to bulge.

    d.   closely spaced valves within the vessels.

    e.   aggregations of B cells and C cells.

7.   The most superior branch of the abdominal aorta, which supplies the liver, stomach, esophagus, gallbladder, duodenum, pancreas, and spleen is:

    a.   the superior mesenteric artery.

    b.   the common iliac artery.

    c.   the gonadal artery.

    d.   the celiac trunk.

    e.   the inferior mesenteric artery.

8.   Which is the *least common* type of white blood cell?

    a.   monocyte

    b.   lymphocyte

    c.   neutrophil

    d.   basophil

    e.   eosinophil

9.   Which of the following applies to elastic arteries?

    a.   They have a poorly defined tunica externa.

    b.   These vessels distribute blood to the skeletal muscles and internal organs of the body.

    c.   The tunica media of these arteries contain a high density of elastic fibers and relatively few smooth muscle cells.

    d.   The tunica media consists of scattered smooth muscle fibers that do not form a complete layer.

    e.   None of the above applies to elastic arteries.

10.   The right common carotid artery and right subclavian arteries are branches of the:

    a.   thyrocervical trunk.

    b.   aortic arch.

    c.   brachiocephalic trunk.

    d.   vertebral artery.

    e.   None of the above.

11.   After leaving the thoracic cavity and passing over the outer border of the first rib, the subclavian artery becomes the:

    a.   axillary artery.

    b.   brachial artery.

    c.   radial artery.

    d.   clavicular artery.

    e.   ulnar artery.

12. Which blood cell lacks a nucleus?
    a.   mature RBC
    b.   immature RBC
    c.   lymphocyte
    d.   mature basophil
    e.   megakaryocyte

13. Capillaries whose endothelial cells are connected by tight junctions are called:
    a.   sinusoidal capillaries.
    b.   continuous capillaries.
    c.   fenestrated capillaries.
    d.   lacteals.
    e.   sinusoids.

14. An artery can be distinguished from a vein by all of the following means *except*:
    a.   arterial walls are thicker than those of veins.
    b.   arterial walls retain their round shape due to their thick and strong walls, while veins tend to collapse because their walls are thinner.
    c.   in cross section, the lumen of a vein appears to be thrown into folds, while that of an artery is smooth.
    d.   arteries have a smaller lumen than does a similar-sized vein.
    e.   All of the above are true.

15. Which of these is *not* a part of a lymph node?
    a.   nodule
    b.   trabeculae
    c.   medulla
    d.   cortex
    e.   white pulp

16. Muscular arteries:
    a.   have a thick tunica media with a large amount of smooth muscle fiber within them.
    b.   are larger than elastic arteries and smaller than arterioles.
    c.   are exemplified by the external carotid, brachial, and femoral arteries.
    d.   are a cushion to prevent rises in blood pressure from causing circulation problems.
    e.   have the attributes of A and C.

17. Which artery delivers blood to most of the large intestine?
    a.   inferior mesenteric artery   d.   superior mesenteric artery
    b.   splenic artery   e.   common hepatic artery
    c.   celiac trunk

18. Which layer of blood vessels consists of collagen and elastic fibers, with cells and fibers running longitudinally?
    a. tunica media
    b. tunica externa
    c. tunica intima
    d. tunica interna
    e. tunica elastica

19. Metarterioles are structural intermediates between which of the following?
    a. arterioles and capillaries
    b. arterioles and venules
    c. venules and capillaries
    d. thoroughfare channels and venules
    e. capillaries and venules

20. Which of the following is defined by the "lub" sound during the cardiac cycle?
    a. closing of semilunar valves at the end of ventricular systole
    b. opening of semilunar valves at the end of ventricular systole
    c. closing of atrioventricular valves at the start of ventricular systole
    d. closing of atrioventricular valves at the end of ventricular systole
    e. opening of atrioventricular valves at the start of atrial systole

21. The ability of leukocytes to move through vessel walls is called:
    a. leukopenia.
    b. chemotaxis.
    c. lymphopenia.
    d. diapedesis.
    e. leukocytosis.

22. Which of the following is the primary source of plasma proteins?
    a. liver
    b. spleen
    c. bone marrow
    d. pancreas
    e. thymus

23. An excessive platelet count ($\geq 1,000,000$ per µl), indicating accelerated platelet formation in response to infection, inflammation, or cancer, is called:
    a. leukocytosis.
    b. thrombocytosis.
    c. hemostasis.
    d. thrombocytopenia.
    e. epistaxis.

24. Normal erythropoiesis in myeloid tissues requires adequate supplies of amino acids, iron, and:

    a. vitamin B_{12}.

    b. vitamin D.

    c. vitamin A.

    d. retinoic acid.

    e. fibrinogen.

Chapter 22 – THE RESPIRATORY SYSTEM

INTRODUCTION

- ### FUNCTIONS of the RESPIRATORY SYSTEM

  - Provides an area for gas exchange between air and circulating blood

  - Supplies the body with oxygen (O_2) and disposes of carbon dioxide (CO_2)

  - Circulates the air between exchange surfaces

  - Protects respiratory surfaces

  - Provides protection against pathogens

  - Provides the ability for speech

  - Helps to regulate blood volume, blood pressure and body fluid pH

- ### FOUR PROCESSES OF RESPIRATION

  - **Pulmonary Ventilation:** air is moved in and out of the lungs

  - **External Respiration:** gas exchange (O_2 loading and CO_2 unloading) between the blood in the capillaries and the air <u>at the lung alveoli</u>

  - **Transport of respiratory gases:** O_2 and CO_2 transported between the lungs and the cells of the body via cardiovascular system, using blood as the transport vehicle

  - **Internal Respiration:** gases exchanged between the systemic blood and the tissue cells <u>at systemic capillaries</u> (O_2 unloading and CO_2 loading)

> **Both the respiratory and cardiovascular systems are irreversibly linked and involved in respiratory functioning.**

279

- ### FUNCTIONAL ANATOMY of the RESPIRATORY SYSTEM

  - **CONDUCTING ZONE:** (*from the nose to the terminal bronchioles*) fairly rigid <u>conduits that carry air</u> to the sites of gas exchange; functions to filter, moisten and warm incoming air

  - **RESPIRATORY ZONE:** (*from the respiratory bronchioles to the alveoli*) <u>actual site of gas exchange in the lungs</u>; structures include *respiratory bronchioles*, *alveolar ducts* and *alveolar sacs* consisting of groups of *alveoli*

ORGANS of the UPPER RESPIRATORY SYSTEM

- The NOSE : the only externally visible part of the respiratory system

  - **External Nose:** composed of bones (frontal, nasal and maxillary bones) and flexible plates of hyaline cartilage

- The NASAL CAVITY (*internal*): from **external nares (nostrils)** to the **internal nares**, or *choanae* (posterior nasal apertures)

  - Divided into right and left halves by the **NASAL SEPTUM**

    - **ANTERIOR PORTION** of the nasal septum is formed of <u>hyaline cartilage</u>.

    - **BONY PORTION** of the nasal septum is formed by the <u>fusion</u> of the perpendicular plate of the <u>ethmoid</u> and the plate of the <u>vomer</u>.

  - Posteriorly continuous with the *nasopharynx* through the *choanae*

  - **Bony boundaries** (lateral and superior walls and the floor) **of the nasal cavity** are formed by the *maxillae*, *nasal* and *frontal bones*, the *ethmoid* and *sphenoid* bones, and by the *hard* and *soft palates*.

  - <u>**NASAL VESTIBULE:**</u> the entryway of the nose, which is protected by hairs that screen out large particles

  - **OLFACTORY MUCOSA:** lines area near the roof of the cavity

  - **RESPIRATORY MUCOSA:** lines the vast majority of the nasal cavity

    - Comprised of **ciliated pseudostratified columnar epithelium** with scattered goblet cells and an underlying lamina propria

- o Richly supplied with sensory nerve endings, rich plexuses of capillaries and thin-walled veins in the lamina propria, which warm incoming air

- o *Sneeze reflex* occurs when irritants contact the mucosa.

- **NASAL CONCHAE** (singular = *concha*), or **turbinate bones**

  - o Consists of 3 mucosa-covered, scroll-like structures projecting medially from each lateral wall of the nasal cavity

  - o **SUPERIOR / MIDDLE / INFERIOR CONCHAE**

  - o **Superior / Middle / Inferior MEATUS**: narrow groove inferior to each concha, appropriately named according to their location

  - o **FUNCTIONS of the CONCHAE and MEATUSES:**

    - Increase air turbulence

    - Together with the nasal mucosa, they filter, moisten and warm air during inhalation.

    - Reclaim the heat and moisture during exhalation

- **HARD PALATE**: forms the bony (anterior) floor of the nasal cavity formed by the maxillary and palatine bones, separating the oral and nasal cavities

- **SOFT PALATE**: forms the fleshy (posterior to hard palate) floor of the nasal cavity, marking the boundary between the superior nasopharynx and the rest of the pharynx

- | The PARANASAL SINUSES |

  - o Ring of air-filled cavities surrounding the nasal cavity

  - o Located in the frontal, sphenoid, ethmoid and maxillary bones; and named according to their location

  - o FUNCTION: **act as resonance chambers in speech**

  - o Lined by the same nasal mucosa and perform the same air-processing functions

- **THE PHARYNX** (*throat*)

  - Funnel-shaped passageway that connects the nasal cavity and the mouth superiorly to the larynx, and inferiorly to the esophagus

  - FUNCTION: **serves as common passageway for both food and air**

  - Consists of skeletal muscle throughout its length, but the nature of the mucosa lining it varies

  - Within the mucosa are **TONSILS**, which destroy pathogens

  3 pharyngeal regions: (*superior to inferior*)

  - <u>**NASOPHARYNX:**</u> (*ciliated pseudostratified columnar epithelial lining*)

    o Located directly posterior to the nasal cavity's CHOANAE (posterior nasal aperture)

    o **AIR passageway only**

  - <u>**OROPHARYNX:**</u> (*thick, protective stratified squamous epithelial lining* due to increased friction and chemical trauma from food and air)

    o Located posterior to the oral cavity

    o **FAUCES**: arch-like entranceway directly behind the mouth

    o **FOOD and AIR passageway**

  - <u>**LARYNGOPHARYNX:**</u> (*stratified squamous epithelial lining*)

    o Located directly posterior to the epiglottis and larynx, and is continuous with both the esophagus and larynx

    o **Common passageway for FOOD and AIR**

ORGANS of the LOWER RESPIRATORY SYSTEM

- **THE LARYNX** (*voice box*)

  - <u>Superiorly</u>, it attaches to the hyoid bone and opens into the laryngopharynx; <u>inferiorly</u>, it is continuous with the trachea.

- **FUNCTIONS of the LARYNX:**
  - ○ Voice production

  - ○ Provides an open airway

  - ○ Acts as a switching mechanism to route air and food into proper channels

- **STRUCTURE of the LARYNX:** Nine (9) LARYNGEAL CARTILAGES connected by membranes and ligaments

  > *ALL, EXCEPT the EPIGLOTTIS, are comprised of HYALINE CARTILAGE.

  - ○ **THYROID CARTILAGE** (1): shaped like an upright open book or a *shield*

    - The largest laryngeal cartilage
    - Forms most of the anterior and lateral walls of the larynx
    - In males, the **laryngeal prominence** (*"book's spine"*) is a thick ridge on the anterior surface of the cartilage, and is commonly called the *"Adam's apple."*

  - ○ **CRICOID CARTILAGE** (1): *"signet-ring-shaped"* cartilage that forms a complete ring of cartilage with greatly expanded posterior portion

    - Located inferior to the thyroid cartilage, and perched on top of the trachea
    - Along with the thyroid cartilage, it protects the glottis and the entrance to the trachea
    - Their broad surfaces provide sites for the attachment of important laryngeal muscles and ligaments

  - ○ **ARYTENOID CARTILAGES** (2): *"ladle-shaped"* cartilages that anchor the vocal cords; articulate with the superior border of the enlarged portion of the cricoid cartilage

  - ○ **CORNICULATE CARTILAGES** (2): *" horn-shaped"* cartilages that articulate with the arytenoid cartilages with which they play a role in the opening and closing of the glottis and the production of sound

  - ○ **CUNEIFORM CARTILAGES** (2): *"wedge-shaped"* cartilages that lie within the *aryepiglottic fold* that extends between the lateral aspect of each arytenoids cartilage ad epiglottis

  - ○ ***EPIGLOTTIS** (1): *"shoehorn-shaped"* laryngeal cartilage

    - Comprised of ELASTIC CARTILAGE

    - Almost entirely covered by mucosa

- Projects superior to the **glottis** (*narrow opening through which inspired, or inhaled, air leaves the pharynx to get to the larynx*)

- Tips inferiorly to cover and seal the laryngeal inlet during swallowing

- Functions in the **cough reflex** in which substances (other than air in the larynx) are expelled

- **THE TRACHEA** (*windpipe*)

  - Descends from the larynx through the neck and into the MEDIASTINUM (thoracic area between the lungs)

  - Ends by branching into the two main bronchi (primary, or 1º bronchi), the RIGHT MAIN BRONCHUS and the LEFT MAIN BRONCHUS, in the mid-thorax

  - **TRACHEAL WALL:** 16-20 C-shaped rings of <u>HYALINE cartilage</u> joined to one another by intervening membranes of fibroelastic connective tissue, called ANNULAR LIGAMENTS

    - The tracheal cartilages stiffen the tracheal walls and protect the airway.

    - Provides flexibility but also prevents trachea from collapsing or overexpanding as pressures change in the respiratory system

    - Keeps the airway open despite pressure changes during breathing

    - **TRACHEAL WALL LAYERS** :

      1. **MUCOUS MEMBRANE, or MUCOSA** (ciliated pseudostratified columnar epithelium + lamina propria)

      2. **SUBMUCOSA** (another layer of connective tissue containing seromucous glands)

      3. **ADVENTITIA** (connective tissue containing the tracheal cartilages)

  - **TRACHEALIS MUSCLE:** smooth muscle fibers, along with soft connective tissue in open posterior regions

    - Functions in decreasing the diameter of trachea

  - **CARINA:** the ridge on internal aspect of the last tracheal cartilage at the point where the trachea branches into main bronchi

    - The *cough reflex* is often initiated here.

- ☐ **THE BRONCHI (singular = bronchus) and their SUBDIVISIONS** :   THE   BRONCHIAL
  TREE

  - The bronchial tree is a <u>system of respiratory passages that branches extensively within the lungs</u>.

  - The right (R) and left (L) main bronchi (1º bronchi) are the largest conduits in the bronchial tree.

  - **<u>Subdivisions</u>:**

    - The R and L MAIN BRONCHI (1º bronchi) supply the lungs and divide into;

    - LOBAR BRONCHI (2º bronchi), which supply the lung lobes and divide into;

    - SEGMENTAL BRONCHI (3º bronchi), which supply the bronchopulmonary segments and through many orders of branching divide into (~20 more orders of branching);

    - BRONCHIOLES/TERMINAL BRONCHIOLES (<1mm to <0.5 mm diameter), which lead into the respiratory zone.

    - The **RESPIRATORY ZONE** consists of structures in decreasing size: **RESPIRATORY BRONCHIOLES – ALVEOLAR DUCTS – ALVEOLAR SACS – ALVEOLI.**

  - Tissue composition of the wall of each main bronchus mimics that of the trachea; changes occur as the conducting tubes decrease in diameter.

    - **Cartilage changes**: cartilage rings are replaced by irregular plates of cartilage as the main bronchi enter the lungs; cartilage is absent in bronchioles but elastin is present

    - **Epithelium changes**: mucosal epithelial lining thins as it changes from *ciliated pseudostratified columnar* to *simple columnar* to *simple cuboidal* in the smallest bronchioles, where cilia or mucus-producing cells are absent

    - **Smooth muscle becomes important**: complete layer of smooth muscle 1st appears in walls of large bronchi and is present throughout the smaller bronchi and bronchioles

- ☐ **ALVEOLI (singular = *alveolus*)** : microscopic air-exchange chambers of the lungs

  - Alveoli are surrounded by a delicate network of pulmonary capillaries and fine elastic fibers.

- **TYPE I cells**: single layer of *squamous epithelial cells*, which lines the walls of alveoli; surrounded by delicate basal lamina

- **TYPE II cells**: *cuboidal epithelial cells*, which secrete a fluid that coats the internal alveolar surfaces and contains **SURFACTANT** (detergent-like); surfactant prevents alveolar walls from sticking together during exhalation

- **ALVEOLAR MACROPHAGES** (*dust cells*): exist in the air space of alveoli and remove the tiniest inhaled particles that are not trapped by the nasal mucosa

- **THE LUNGS and the PLEURAE** (singular = *pleura*)

  - **The Pleurae** : double-layered flattened sacs, around each lung, whose walls consist of serous membranes

    o **PARIETAL PLEURA**: (outer layer) covers internal surface of thoracic wall, superior surface of diaphragm and lateral surfaces of mediastinum

    o **VISCERAL PLEURA**: (inner layer) covers external surface of the lung

    o **PLEURAL CAVITY**: potential space (between the parietal and visceral pleurae) filled with pleural fluid, which decreases friction between the lungs and the thoracic wall during breathing

  - **Gross Anatomy of the Lungs**

    o Paired R and L lungs and their pleural sacs; each lung is roughly cone-shaped.

    o The lungs consist mainly of air tubes and spaces and the balance of its tissue, called **STROMA**, which is a framework of connective tissue containing many elastic fibers.

    o Therefore, the lungs are light, soft, spongy, elastic organs.

    o The anterior, lateral and posterior surfaces of a lung contact the ribs and form a continuously curving **COSTAL SURFACE**.

    o **APEX**: rounded, superior tip

o **BASE**: concave inferior surface that rests on the diaphragm

o **HILUS**: indentation on the medial surface of each lung through which blood vessels, bronchi, lymph vessels and nerves enter and exit the lung

o **LEFT LUNG**: somewhat smaller

- Contains the **CARDIAC NOTCH** (deviation in its anterior border that accommodates the heart)

- Divided into 2 lobes (**upper and lower**, or **superior and inferior**) by the **OBLIQUE FISSURE**

o **RIGHT LUNG**:

- Divided into 3 lobes (**upper** or superior, **middle**, and **lower** or inferior lobes) by the **OBLIQUE FISSURE** and the **HORIZONTAL FISSURE**

o Each lung lobe is further subdivided into 10 **BRONCHOPULMONARY SEGMENTS**, which are separated by thin partitions of dense connective tissue.

- The segments **limit the spread of some diseases** within the lung.

■ | **BLOOD SUPPLY and INNERVATION of the lungs** |

o **PULMONARY ARTERIES** (branch along with bronchial tree) deliver O_2-poor blood, from the heart to the lungs for oxygenation.

o **PULMONARY CAPILLARY** NETWORK surrounds the alveoli.

o **PULMONARY VEINS** carry oxygenated blood from alveoli of the lungs to the heart.

o **SYMPATHETIC NERVE fibers** dilate the air tubes.

o **PARASYMPATHETIC NERVE fibers** constrict air tubes.

o **VISCERAL SENSORY fibers** and the other nerve fibers enter each lung through the **PULMONARY PLEXUS** on the root of the lung.

PULMONARY VENTILATION

- The **MECHANISM of breathing**, or *pulmonary ventilation*, consists of 2 phases, INSPIRATION and EXPIRATION.

  - **INSPIRATION** (inhalation) : period when **air flows into the lungs**, causing increased volume and decreased pressure within the thoracic cavity via the **inspiratory muscles, DIAPHRAGM and INTERCOSTALS (muscles of the ribs)**

    - **Action of the DIAPHRAGM:** contraction causes it to move inferiorly and to flatten; results in <u>increased vertical (height) dimension of the thoracic cavity</u>

    - **Action of the INTERCOSTALS:** contraction causes the ribs to raise; results in <u>increased right-to-left dimension ("*lateral expansion*") of the thoracic cavity</u>

  - **EXPIRATION** (exhalation) : period when **gases exit the lungs**

    - Mainly a passive process in which the inspiratory muscles relax

    - Decreased volumes of the thorax and lungs occur simultaneously

NEURAL CONTROL of PULMONARY VENTILATION

  - Main respiratory center is located in the reticular formation of the medulla oblongata = **RVLM (rostral ventrolateral medulla)**

  - RVLM is considered as a **pacemaker** whose neurons generate the **basic ventilatory rhythm and rate**.

  - Basic pattern can be modified by higher centers of the central nervous system:

    - The limbic system and the hypothalamus confer emotional influence.

    - The cerebral cortex mediates conscious control.

  - The basic pattern is also modified by **chemoreceptors**, which sense the chemistry of blood.

Chapter 23 – THE DIGESTIVE SYSTEM

OVERVIEW : Two (2) groups of digestive system organs

- **The ALIMENTARY CANAL** (*GI Tract,* or *gastrointestinal tract*)

  - Structures include the **mouth, pharynx, esophagus, stomach, small intestine, large intestine, rectum, anal canal and anus**

  - In a cadaver, the GI tract is about 9 meters (30 feet) long; but shorter in a living person due to presence of muscle tone.

- **ACCESSORY DIGESTIVE ORGANS**

  - Include the **teeth, gallbladder**, and **various large digestive glands** (*which secrete saliva, digestive enzymes and bile*), such as the **salivary glands, liver,** and **pancreas**

  - These structures lie external to and connect to the GI tract via ducts.

- **DIGESTIVE PROCESSES** (6 essential activities)

  1. Ingestion
  2. Mechanical Digestion – chewing, churning, segmentation
  3. Propulsion (movement of food through GI tract)
     - **Peristalsis** – alternate waves of contraction and relaxation of musculature in the organ walls
  4. Chemical Digestion – enzymes secreted by digestive glands into the lumen of the GI tract
  5. Absorption – transport of digested end products from the lumen to blood and lymph capillaries
  6. Defecation – elimination of indigestible substances from the body as feces

ANATOMY OF THE DIGESTIVE SYSTEM

- **PERITONEAL CAVITY and the PERITONEUM**

  - **PERITONEUM:** the serous membrane in the abdominopelvic cavity, which has a PARIETAL LAYER (on the internal surface of the body wall) and a VISCERAL LAYER (on the viscera)

  - **PERITONEAL CAVITY** contains serous fluid, which decreases friction as the organs move

- **HISTOLOGY** of the ALIMENTARY (GI Tract) CANAL WALL

  - **FOUR TISSUE LAYERS of the walls** (*from the lumen – outward*):

    - **MUCOSA** (or *mucous membrane*) : three sublayers

      - **Epithelial lining** – absorbs nutrients and secretes mucous

      - **Lamina propria** – loose areolar or reticular connective tissue whose capillaries nourish the epithelium and absorb digested nutrients
        - Contains **MALT** (*mucosa-associated lymphoid tissue*)

      - **Muscularis mucosae** – thin layer of smooth muscle
        - Produces local movements of the mucosa

    - **SUBMUCOSA**

      - Layer of connective tissue containing major blood and lymphatic vessels and nerve fibers

      - Intermediate between loose areolar and dense irregular connective tissue

    - **MUSCULARIS EXTERNA** : comprised of two layers of smooth muscle, which is responsible for *peristalsis* and *segmentation*

      - **Circular muscle layer** (*inner*) – muscle fibers orient around the circumference of the canal; *squeezes* the alimentary tube

      - **Longitudinal muscle layer** (*outer*) – fibers orient along the length; *lengthens* the alimentary tube

    - **SEROSA** (*visceral peritoneum*)

      - **Simple squamous epithelium** (*outer* mesothelium)

      - **Areolar connective tissue** (*inner*)

      - Parts of the GI tract not associated with the peritoneal cavity lack a **serosa** and have an **adventitia** (ordinary fibrous connective tissue)

- ▪ **NERVE PLEXUSES** : (singular = *plexus*)

  - ○ Networks of nerve fibers present in the tissue layers of the alimentary canal wall

  - ○ **MYENTERIC NERVE PLEXUS** – innervates *muscularis externa* between the circular and longitudinal layers
    - • Controls peristalsis and segmentation

  - ○ **SUBMUCOSAL NERVE PLEXUS** – extends inward and lies within the submucosa
    - • *Signals the glands in the mucosa* to secrete and the *muscularis mucosa* to contract

- • **The MOUTH and ASSOCIATED ORGANS**

  - ▪ The **MOUTH** (*oral cavity*) – mucosa-lined

    - ○ **Regions and structures:** vestibule, oral cavity proper, lips (*labia*), cheeks, hard palate (anterior), and soft palate (posterior)

    - ○ **Histology**

      - • Internal mucosa (epithelium and lamina propria only)

      - • Thin submucosa

      - • External layer of muscle or bone

      - • <u>Lining of the mouth:</u> **thick, stratified squamous epithelium** (slight keratinization in the lining of the tongue, palate, lips and gums)

  - ▪ The **TONGUE**

    - ○ Muscle constructed of **interlacing fascicles of skeletal muscle** fibers

    - ○ INTRINSIC MUSCLES change the shape of the tongue but not its position.

    - ○ EXTRINSIC MUSCLES alter the position of the tongue.

  - ▪ The **SALIVARY GLANDS**

    - ○ Compound tubuloalveolar glands

o The glands' *secretory cells* are **serous cells** (produce watery secretion containing enzymes and ions of saliva) and **mucous cells** (produce **mucus**).

o **FUNCTIONS:**

- Moistens mouth

- Dissolves food chemicals so that they can be tasted

- Wets or moistens food

- Binds food together into a **bolus**

- Enzymes begin digestion of starches

- Bicarbonate buffer in saliva neutralizes acids produced by oral bacteria

- Saliva also contains bactericidal enzymes, antiviral substances, antibodies and a cyanide compound

o | **Small INTRINSIC Salivary Glands** |

- Scattered with the mucosa of the tongue, palate, lips and cheeks

- Saliva from these glands keep mouth moist at all times

o | **EXTRINSIC Salivary Glands** |

- Lie external to the mouth but connect to it via ducts

- Secrete saliva only during eating or in anticipation of a meal

- **PAROTID GLAND:** the largest salivary gland

  ▪ Lies anterior to the ear

  ▪ **Parotid duct** opens into the mouth

- **SUBMANDIBULAR GLAND (submaxillary):**

  ▪ Submandibular duct opens directly lateral to the tongue

- **SUBLINGUAL GLAND**:

  - 10-12 ducts open into the mouth, directly superior to the gland

- The ⬚ **TEETH** ⬚

  o Lie in sockets (**alveoli**) in the gum-covered margins of the mandible and maxilla

  o **Mastication:** tear and grind the food, breaking it into smaller fragments

  o <u>**TOOTH STRUCTURE:**</u>

    - **CROWN** – exposed and covered by 2.5 mm thick layer of enamel

    - **ROOT(s)** – located in the socket(s)

- The ⬚ **PHARYNX** ⬚ : oropharynx and laryngopharynx
  (*More focus on this structure in the Respiratory System section*)

  o Passageways for food, fluids and inhaled air

  o **Pharyngeal constrictors** are skeletal muscles

- The ⬚ **ESOPHAGUS** ⬚

  o Muscular tube that **propels swallowed food to the stomach**

  o Has all four tissue layers of the alimentary canal

  o Begins as a continuation of the pharynx in the mid-neck region, descends through the thorax on the anterior surface of the vertebral column and passes through the **esophageal hiatus** in the diaphragm to enter the abdomen

  o **Cardiac orifice:** junction at which the esophagus joins the stomach

  o A **cardiac sphincter** acts to close off the lumen and prevent regurgitation of acidic stomach juices into the esophagus.

  o Non-keratinized stratified squamous epithelium lining the junction of the esophagus and stomach

o ☐ The ⅓-⅓-⅓ **RULE of the ESOPHAGEAL MUSCLES** ☐ involves the MUSCULARIS EXTERNA LAYER, which consists of the following:

- *Skeletal muscle* in the **SUPERIOR ⅓ of the esophagus**
- A *mixture* of skeletal and smooth muscle in the **MIDDLE ⅓**
- *Smooth muscle* in the **INFERIOR ⅓**

- ☐ **THE STOMACH** ☐

  ■ J-shaped; widest part of the GI tract

  ■ Temporary storage tank in which food is churned and turned into a paste called **CHYME**

  ■ Pepsin and HCl (*hydrochloric acid*) secreted to start breakdown of food proteins

  ■ Lies in the superior left of the peritoneal cavity; left hypochondriac, epigastric and umbilical regions of the abdomen, directly inferior to the diaphragm and anterior to the spleen and pancreas

  ■ **GROSS ANATOMY of the STOMACH:**

    o **CARDIAC REGION (or *cardia*)** – ring-shaped zone encircling cardiac orifice

    o **FUNDUS** – the stomach's dome tucked under the diaphragm

    o **BODY** – large middle portion of the stomach

    o **PYLORIC REGION** (funnel-shaped) –

      - Pyloric ANTRUM

      - Pyloric CANAL

    o **PYLORUS** – terminus of stomach

      - **Pyloric sphincter** controls entry of **chyme** into the intestine

    o **GREATER CURVATURE** – convex left surface of the stomach, which opens into the **greater omentum** o **LESSER CURVATURE** – concave right margin of the stomach, which opens into the **lesser omentum**

o **RUGAE of mucosa** – numerous longitudinal folds on internal surface of empty stomach, which flattens as the stomach fills (1.5 L - 4 L capacity)

o **OBLIQUE LAYER of the muscularis externa** is only present in the walls of the stomach; more specifically, deep to the circular layer of the *muscularis externa*

■ **MICROSCOPIC ANATOMY of the STOMACH**

o The lining epithelium of the stomach is comprised of *simple columnar cells.*

o This layer contains <u>increased amounts of goblet cells,</u> which secrete bicarbonate-buffered *mucus* that protects the stomach wall from pepsin and HCl.

• **THE SMALL INTESTINE (SI)**

■ Longest part of the GI tract (alimentary canal)

■ **Site of most enzymatic digestion and virtually all absorption of nutrients**

■ <u>Three subdivisions:</u>

o **DUODENUM** (~5% of SI length):

  • Receives digestive enzymes from the pancreas via the pancreatic duct

  • Receives bile from the liver and gallbladder via the bile duct

  • Main pancreatic duct and bile duct join to form the **hepatopancreatic ampulla and sphincter**

  • Ampulla opens into the duodenum at the **major duodenal papilla**

o **JEJUNUM** (~40%): superior left

o **ILEUM** (~55%): inferior right

> ***Mnemonic Device:*** ***"Don't Jiggle It"*** --- to remember the three subdivisions of the small intestine (**D**uodenum – **J**ejunum – **I**leum), in order from the proximal to distal (or superior to inferior) regions.

- | Three structural modifications of wall of the SI that amplify its absorptive surface | :

  o **CIRCULAR FOLDS** or ***plicae circulares*** – permanent, transverse ridges (~1 cm tall) of the mucosa and submucosa

  - Increase the surface area

  - The folds force the chyme to spiral through the lumen to increase absorption time.

  o **VILLI** – finger-like projections of the mucosa (>1 mm tall)

  - Villi are covered by a *simple columnar epithelium* made up primarily of **absorptive cells** (***enterocytes***) specialized for absorbing digested nutrients

  - Digested fats enter the **lacteals** (lymph capillaries)

  - All end products of nutrient digestion enter the blood capillaries.

  o **MICROVILLI** – found on apical surfaces of the **absorptive cells** (***enterocytes***)

  - Increase absorptive surfaces

  - Plasma membrane contains enzymes that complete the final stages of breakdown of nutrient molecules

- | **HISTOLOGY of the WALL of the Small Intestine (SI)** | :

  o All typical layers of the GI tract occur in the SI.

  o The lining epithelium occurs on the villi and the surface between the villi.

  o **Absorptive cells** (***enterocytes***) contain high number of mitochondria, goblet cells and **enteroendocrine cells**.

  o **Intestinal crypts** (***intestinal glands*** or ***crypts of Lieberkuhn***) are mucosal tubes between the villi.

- Epithelial cells lining these crypts **secrete intestinal juice**, a watery liquid that mixes with chyme in the intestinal lumen

- Inner epithelia are completely renewed every 3-6 days due to the destructive effects of the digestive enzymes.

- ## THE LARGE INTESTINE (LI)

  - The last major organ in the GI tract

  - The material that reaches it is a largely digested residue that contains few nutrients.
    - Residue remains in the LI for ~12-24 hours

  - Little additional breakdown of food occurs here.

  - **MAIN FUNCTION:** absorb water and electrolytes from the digested mass resulting in semisolid **feces**

  - Propulsion is sluggish and weak; **mass peristaltic movements**

  - Three SPECIAL FEATURES of the LARGE INTESTINE (LI):

    - **TENIAE COLI** – consists of three (3) longitudinal strips, spaced at equal intervals around the circumference of the **cecum** and the **colon**

      - Function in maintaining muscle tone and providing support

      - As a result, the teniae coli cause the LI to pucker into sacs.

    - **HAUSTRA** (singular = *haustrum*) – sacs formed by teniae coli

    - **EPIPLOIC APPENDAGES** (*omental appendices*) – fat-filled pouches of visceral peritoneum hanging from the LI

      - Unknown function

  - Five SUBDIVISIONS of the LARGE INTESTINE(LI):

    - **CECUM** – located in the right iliac fossa

- Forms the junction at which the ileum of the small intestine opens into the large intestine

- **ILEOCECAL VALVE** controls chyme entering the LI

○ **VERMIFORM APPENDIX** – blind tube that opens into the posteromedial wall of the cecum

- Contains large masses of lymphoid tissue in its wall

- Probably functions in gathering antigens and neutralizing harmful pathogens

○ **COLON** – distinct segments

- Ascending colon – right colic flexure (or hepatic flexure) at the level of the right kidney

- Transverse colon – left colic flexure (or splenic flexure) directly anterior to the spleen

- Descending colon

- Sigmoid colon (S-shaped)

○ **RECTUM** – no teniae coli present

- Its longitudinal muscle layer is complete and well developed.

○ **ANAL CANAL** – ~3 cm long

- Includes the external anal sphincter

- | MICROSCOPIC ANATOMY of the LARGE INTESTINE (LI) | :

○ Villi are absent in the LI.

○ Intestinal crypts are present.

○ Goblet cells are more abundant – for increased mucous secretion.

○ Absorptive cells (or "columnar cells") of the LI take in water and electrolytes.

- | **THE LIVER** |

    - The largest gland in the body

    - <u>**DIGESTIVE FUNCTION:**</u> produce **bile** to break down fats in the SI

    - <u>**METABOLIC FUNCTIONS:**</u> (carried out by hepatocytes or liver cells)

        o Picks up glucose from blood returning from GI tract and stores it as glycogen

        o Processes fats and amino acids

        o Stores certain vitamins

        o **Detoxifies many poisons and drugs in the blood**

        o Makes the blood proteins

    - | Two hepatic surfaces | :

        o **DIAPHRAGMATIC SURFACE** – the superior, anterior, and posterior surfaces of the liver, which directly and indirectly contact the diaphragm

        o **VISCERAL SURFACE** – the inferior surface of the liver

        o The **BARE AREA** (superior part) is fused to the diaphragm and devoid of peritoneum

        o **Four lobes (right and left lobes, caudate and quadrate lobes) are divided by the following:**

            • The ***falciform ligament*** (a ventral mesentery) *on the diaphragmatic surface* and the ***fissure on the ventral surface*** **mark the division between the right and left lobes of the liver.**

            • <u>On the posterior (*dorsal*) surface</u> of the liver, the *impression left by the inferior vena cava* (IVC) marks the division between the right lobe and the small **caudate lobe.**

            • The **quadrate lobe** lies <u>inferior to the caudate lobe,</u> sandwiched between the left lobe and the gallbladder.

            > **The new terminology actually subdivides the lobes of the liver into segments** based on the major subdivisions of the hepatic artery proper, hepatic portal vein, and hepatic ducts: **anterior, posterior, medial, and lateral segments**

- <u>**PORTA HEPATIS**</u> – "gateway" to the liver where most of the major vessels (hepatic portal vein and hepatic artery proper) and nerves enter and leave the liver

  o The RIGHT and LEFT HEPATIC DUCTS (**containing bile**) exit the porta hepatis and fuse to form the <u>common hepatic duct</u>.

- MICROSCOPIC ANATOMY of the LIVER :

  o **(LIVER) LOBULE** – the basic functional unit of the liver

    • Millions of these units exist in the liver

    • Hexagonal solid structure

    • Consists of plates of liver cells (*hepatocytes* = specialized epithelial cells) radiating out from a **CENTRAL VEIN**

    • **Six portal triads (*portal areas* or *hepatic triads*) exist in each liver lobule.**

    • The central veins of each lobule ultimately merge to form the RIGHT and LEFT HEPATIC VEINS, which drain blood from the liver and return it to the systemic circuit via the inferior vena cava (IVC).

  o **PORTAL (or HEPATIC) TRIAD** (or *portal tract*) – located at almost every corner of the six-sided liver lobule, it <u>consists of branches of three main vessels</u>:

    • **PORTAL ARTERIOLE** – a branch of the hepatic artery proper

    • **PORTAL VENULE** – a branch of the hepatic portal vein

    • **BILE DUCT**– a branch of the hepatic (bile) duct

      ▪ **BILE CANALICULI** carry bile to **BILE DUCTULES** that lead to the portal triads.

      ▪ BILE DUCTS from each lobule unite to form the **RIGHT and LEFT HEPATIC DUCTS.**

    • Arterial blood supplies hepatocytes with oxygen.

    • Venous blood delivers substances from the intestines, for processing by the hepatocytes.

    o **KUPFFER CELLS** (*hepatic macrophages* or *stellate reticuloendothelial cells*) – star-shaped cells found inside the **sinusoids**, which <u>destroy bacteria and other foreign particles</u> in the blood

- **THE GALLBLADDER**

  - A muscular sac resting in a shallow depression on the visceral surface of the liver

  - Divided into three regions: the FUNDUS, the BODY, and the NECK

  - FUNCTION: **stores and concentrates (or modifies) bile** that is produced by the liver

  - Honeycomb pattern of mucosal foldings internally enables expansion.

  - The **CYSTIC DUCT** leads from the gallbladder toward the *porta hepatis*:

    o Joins the **COMMON HEPATIC DUCT** (from the liver) to form the **COMMON BILE DUCT**, which empties into the duodenum

  - At the duodenum, a muscular **HEPATOPANCREATIC SPHINCTER** (*sphincter of Oddi*) surrounds the lumen of the common bile duct and the **duodenal ampulla.**

  - The **DUODENAL AMPULLA** opens into the duodenum at the **HEPATODUODENAL PAPILLA** (or *duodenal papilla*), which is a small raised projection.

  - **Contraction of the hepatopancreatic sphincter** seals off the passageway and **prevents bile from entering the small intestine.**

- **THE PANCREAS** (an *exocrine* and *endocrine gland*)

  - **EXOCRINE FUNCTION:** produces most of the digestive enzymes and buffers in the SI (small intestine)

  - **ENDOCRINE FUNCTION:** produces hormones (insulin and glucagon) that regulate the levels of sugar in the blood

  - Tadpole-shaped with three regions:

    o The HEAD of the pancreas is broad and lies within the loop formed by the duodenum as it leaves the pylorus.

o The BODY region is slender and extends transversely toward the spleen.

o The TAIL region is short and bluntly rounded.

- The **(MAIN) PANCREATIC DUCT** (or *duct of Wirsung*) extends through the length of the pancreas:

  o Joins the **bile duct** to form the **HEPATOPANCREATIC AMPULLA**, which empties into the duodenum at the **major** (or **greater**) **duodenal papilla**

- **ACCESSORY PANCREATIC DUCT** (or *duct of Santorini*) lies in the head of the pancreas and drains into the duodenum at a separate papilla, the **lesser duodenal papilla**.

- **BLOOD SUPPLY** :

  o The **PANCREATIC ARTERIES and PANCREATICODUODENAL ARTERIES** (**superior** and **inferior**) are major branches from the *splenic, superior mesenteric,* and *common hepatic arteries,* whose branches through which arterial blood reaches the pancreas.

  o The **SPLENIC VEIN** and its branches drain the pancreas.

- **HISTOLOGY of the PANCREAS** :

  o Partitions of connective tissue divide the pancreatic tissue into distinct lobules.

  o The pancreas is a compound *tubuloacinar gland*.

  o **PANCREATIC ACINI** (singular = *acinus*) : within each lobule, the ducts branch repeatedly before ending in these blind pockets

  - *Simple cuboidal epithelium* lines each pancreatic acinus.

  - DIGESTIVE FUNCTION: They **secrete PANCREATIC JUICE** (a mixture of water, ions, and pancreatic digestive enzymes) into the duodenum.

    - The **pancreatic enzymes** in the pancreatic juice do most of the digestive work in the SI, by breaking down ingested materials into small molecules suitable for absorption.

    - The **PANCREATIC DUCTS secrete buffers** (primarily sodium bicarbonate) in a watery solution, which is important in neutralizing the acid in chyme and stabilizing the pH of the intestinal contents.

- o $\boxed{\textbf{PANCREATIC ISLETS}}$: scattered between the acini

  - Account for only ~1% of the cellular population of the pancreas

  - Perform the endocrine function of the pancreas, by **producing the hormones insulin and glucagon**

- $\boxed{\textbf{THE MESENTERIES}}$

  - **Double-layered sheets of peritoneum** that connect peritoneal organs to the dorsal and ventral body wall

  - **FUNCTIONS of the MESENTERIES:**

    - o Support the abdominal digestive organs

    - o Also store fat and carry blood vessels and nerves

  - Visceral organs that lack a mesentery and are fused to the posterior body wall are called **RETROPERITONEAL ORGANS**

  - **VENTRAL MESENTERIES** (and associated organs/structures):

    - o <u>FALCIFORM LIGAMENT</u> (liver)

      - Binds the anterior aspect of the liver to the anterior abdominal wall and the diaphragm

    - o <u>LESSER OMENTUM</u> (liver)

      - Extends from the fissure of the liver and porta hepatis to the lesser curvature of the stomach

  - **DORSAL MESENTERIES** (and associated organs/structures):

    - o <u>GREATER OMENTUM</u> (stomach)

      - Connects the greater curvature of the stomach to the posterior abdominal wall in a roundabout way

      - Extends inferiorly to cover the transverse colon and the coils of the SI, like a butterfly net

o   <u>MESENTERY PROPER</u> (ileum and jejunum)

o   <u>TRANSVERSE MESOCOLON</u> (transverse colon)

o   <u>SIGMOID MESOCOLON</u> (sigmoid colon)

o   <u>LESSER OMENTUM</u> (on ventral surface of the liver)

Chapter 24 – THE URINARY SYSTEM

ORGANS OF THE URINARY SYSTEM

- The **KIDNEYS** (the major excretory organs of the system)

- FUNCTIONS of the KIDNEYS:

    - Cleanse the blood of nitrogenous wastes, toxins, excess ions and water, and other unnecessary or undesirable substances by forming <u>urine</u>

    - The kidneys also maintain proper chemical composition of the blood and other body fluids.

- **Nitrogenous compounds (main waste products) excreted in urine are:**

    - **UREA** – derived from breakdown of amino acids during normal recycling of body's proteins

    - **URIC ACID** – from turnover of nucleic acids

    - **CREATININE** – formed by the breakdown of creatine phosphate, a molecule in muscle that stores energy for the manufacture of ATP

- ## ORGANS for TRANSPORTING and STORING URINE :

    - **URETERS** (tubes that carry urine from the kidneys to the bladder)

    - **URINARY BLADDER** (temporary storage tank for urine)

    - **URETHRA** (tube that carries urine to the body exterior)

THE KIDNEYS

- ## GROSS ANATOMY of the KIDNEYS

    - LOCATION and EXTERNAL APPEARANCE

        - Red-brown, bean-shaped kidneys **lie in the superior lumbar region** of the posterior abdominal wall.

        - The **right kidney is crowded by the liver** and lies slightly inferior to the left kidney.

o CONVEX lateral surface

o CONCAVE medial surface

o Its vertical cleft contains the **RENAL HILUS** where vessels and nerves enter and exit the kidney.

o | **Several layers of supportive tissue surround each kidney** | :

   (*Deep* or *internal* to *superficial* or *external*)

   • **RENAL CAPSULE** (or *fibrous capsule*): dense connective tissue layer of collagen fibers, which adheres directly to the kidney's surface

      ▪ Maintains the shape of the kidney

      ▪ Forms a barrier to inhibit the spread of infection

      ▪ Provides mechanical protection

   • **ADIPOSE CAPSULE** (or *perinephric fat*, or *perirenal fat*): layer of fat around the kidney, which provides cushioning

   • **RENAL FASCIA:** dense outer layer formed by the collagen fibers that extend outward from the inner *renal (fibrous) capsule* through the *adipose* capsule, or *perinephric fat*

      ▪ Anchors the kidney to surrounding structures

      ▪ Posteriorly, it is bound to deep fascia surrounding the muscles of the body wall.

      ▪ Anteriorly, it is attached to the peritoneum and to the *anterior renal fascia* of the opposite side.

   • **PARARENAL FAT:** separates the posterior and lateral portions of the renal fascia from the body wall

▪ | INTERNAL STRUCTURE of the KIDNEYS | :

o **2 distinct regions of kidney tissue:**

   • | **RENAL CORTEX** | : the granular and reddish-brown outer layer of the kidney, which is in contact with the *renal (or fibrous) capsule*

306

- **RENAL COLUMNS** are inward extensions of the cortex, which separate the adjacent pyramids

- **RENAL MEDULLA** : darker region located internal to the renal cortex

  - Consists of **MEDULLARY PYRAMIDS** (or *renal pyramids*), which are distinctly cone-shaped or triangular masses containing striations (fine grooves)

  - The base of each pyramid faces the cortex, and the tip projects into the renal sinus.

  - Contains **RENAL PAPILLAE** (singular = *papilla*), which are the pyramids' tips, or apices (singular = *apex*)

  - **RENAL LOBE:** the area that contains a renal pyramid, the overlying area of renal cortex, and adjacent tissues of the renal columns

- **RENAL SINUS:** large space within the medial part of the kidney opening to the exterior through the renal hilus

- **RENAL PELVIS** (flat, *funnel-shaped* tube): expanded superior part of the ureter, which has branching extensions of 2 or 3 **major calices**

- **MAJOR CALICES** [*calyx* (singular) = "*cup-shaped*"] - further branch out as **minor calices**

- **MINOR CALICES:** smaller *cup-shaped* tubes that enclose the papillae of the pyramids

- The calices **collect urine draining from the papillae** and empty it into the renal pelvis, then into the ureters, and finally into the urinary bladder.

- **GROSS VASCULATURE** : (BLOOD FLOW through the KIDNEY)

  - Under normal resting conditions, ~25% of the heart's systemic output (**AORTA**) reaches the kidneys via the large **RENAL ARTERIES** which <u>branch out into the kidney as follows:</u>

    - **RENAL** ARTERIES
    - **SEGMENTAL** ARTERIES (*enter renal hilus*)
    - **LOBAR** ARTERIES
    - **INTERLOBAR** ARTERIES (*lie in renal columns between medullary pyramids*)
    - **ARCUATE** ARTERIES (*arch over the bases of pyramids*)

- **INTERLOBULAR** ARTERIES (*radiate outward from the arcuate arteries and supply the cortical tissue, and divide the cortical tissue into lobules*)
- **AFFERENT ARTERIOLES**
- **GLOMERULUS** (*tuft of capillaries*)
- **EFFERENT ARTERIOLES** drain blood into:
- **Peritubular capillaries** (*arise from efferent arterioles emptying the cortical glomeruli and surround uriniferous tubules in the interstitial connective tissue of the renal cortex*); **Vasa recta** (*arise from efferent arterioles emptying the juxtamedullary glomeruli and run alongside the loops of Henle in the deepest part of the renal cortex*) -- blood leaves the renal cortex and drains sequentially into:
- **INTERLOBULAR** <u>VEINS</u>
- **ARCUATE** VEINS
- **INTERLOBAR** VEINS
- **RENAL** VEINS
- **INFERIOR VENA CAVA** (IVC)

- ○ >90% of blood entering the kidney perfuses the cortex.

- ○ The **veins** of the kidney essentially trace the pathway of the arteries **in reverse**, but **there are NO LOBAR or SEGMENTAL VEINS.**

<u>SUMMARY of BLOOD FLOW in the KIDNEYS</u>:
(A. = arteries / V. = veins)

RENAL A. → **SEGMENTAL** A. → **LOBAR** A. → **INTERLOBAR** A. → **ARCUATE** A. → **INTERLOBULAR** A. → **AFFERENT ARTERIOLES** → **GLOMERULUS** → **EFFERENT ARTERIOLES** → **PERITUBULAR CAPILLARIES / VASA RECTA** → **INTERLOBULAR** V. → **ARCUATE** V. → **INTERLOBAR** V. → **RENAL** V. → **INFERIOR VENA CAVA**

- ■ NERVE SUPPLY of the KIDNEYS :

  - ○ Provided by the **RENAL PLEXUS**: a network of AUTONOMIC FIBERS and GANGLIA on the RENAL ARTERIES

  - ○ The plexus is supplied by sympathetic fibers, which control the diameters of kidney arteries and influence urine-forming functions of the uriniferous tubules.

- ### MICROSCOPIC ANATOMY of the KIDNEY :

(URINIFEROUS TUBULE comprised of a NEPHRON and a COLLECTING TUBULE, and BLOOD VESSELS)

- **URINIFEROUS TUBULE:** the main structural and *functional unit of the kidney* (>1 million tubules per kidney)

  - **NEPHRON:** <u>urine-forming structure</u>

    - COMPONENTS: renal corpuscle, proximal convoluted tubule, loop of Henle and distal convoluted tubule

  - **COLLECTING TUBULE (DUCT):** concentrates urine by removing water from the urine formed in the nephron

    - Lined by simple epithelium throughout its length

- ### MECHANISMS OF URINE PRODUCTION

  - **FILTRATION:** filtrate of blood leaves the kidney capillaries and enters the nephron

  - **REABSORPTION:** (*passive process*) most of nutrients, water and essential ions are reclaimed from the filtrate and returned to the blood of capillaries

  - **SECRETION:** (*active process*) moves additional undesirable molecules into the collecting tubule from the blood of surrounding capillaries

FOCUS on the COMPONENTS of the URINIFEROUS TUBULE :

- The NEPHRON : (RENAL CORPUSCLE / PCT / LOOP of HENLE / DCT)

  - **<u>RENAL CORPUSCLE</u>** (occur in CORTEX ONLY) consists of:

    - a GLOMERULUS (tuft/ball of fenestrated capillaries) surrounded by a *glomerular capsule*

- **GLOMERULAR CAPSULE** (or *Bowman's capsule*), which has a hollow interior called the **CAPSULAR SPACE** (*separates the parietal and visceral epithelial layers*).

  - **VASCULAR POLE:** the connection between the parietal and visceral epithelia

  - The **glomerular capillaries** are connected to the bloodstream via the **afferent** and **efferent arterioles**.

2 layers of the glomerular capsule :

  - **PARIETAL LAYER** (*simple squamous epithelium*) contributes to structure only

  - **VISCERAL LAYER** clings to the glomerulus

    o Unusual, branching epithelial cells (**PODOCYTES**) with interdigitating foot processes (**PEDICELS**, or *secondary processes*) surround the glomerular capillaries.

- Filtrate passes into the capsular space through **FILTRATION SLITS** (*thin clefts between the pedicels*).

> **The GLOMERULAR CAPILLARIES <u>produce</u> the FILTRATE that moves through the rest of the uriniferous tubules, FORMING URINE.**

- **FILTRATION MEMBRANE or APPARATUS** (*barrier*) :

  - The actual filter that lies between the blood in the glomerulus and the capsular space

  - <u>**This filter has 3 layers of physical barriers:**</u>

    1. *Fenestrated endothelium* of capillary

    2. *Filtration slits* of glomerular epithelia, each covered by a thin slit diaphragm

    3. *Intervening basement membrane* consisting of fused basal laminae of the endothelium and the podocyte epithelium

- ▪ If this dense layer encircles two or more capillaries, **MESANGIAL CELLS** are situated between the endothelial cells of adjacent capillaries.

    - ○ Mesangial cells provide **physical support for capillaries**.

    - ○ They **engulf organic materials** that might otherwise clog the dense layer.

    - ○ They **regulate the diameters of the glomerular capillaries**. Hence, they have a role in the regulation of glomerular blood flow and filtration.

- ○ | **PROXIMAL CONVOLUTED TUBULE (PCT)** | (nephron component that is confined entirely to the renal CORTEX)

    - • Most active in **reabsorption** and **secretion**

    - • *Cuboidal epithelial cells* with luminal surface microvilli

    - • Contains **increased number of mitochondria** to provide energy for reabsorption

- ○ | **LOOP of HENLE (or NEPHRON LOOP)** |

    - • **DESCENDING LIMB:**

        - ▪ 1st part has similar structure to and is continuous with the proximal convoluted tubule (PCT)

        - ▪ 2nd part of the limb is considered the **thin segment**; permeable *simple squamous epithelium*

    - • **ASCENDING LIMB:**

        - ▪ Its **thick segment** begins deep in the medulla and contains active transport mechanisms that pump sodium and chloride ions out of the tubular fluid.

        - ▪ Its **thin segment** is freely permeable to water, but relatively impermeable to ions and other solutes.

    - • <u>Each limb contains a thick segment and a thin segment.</u> (*Thick* and *thin* refer to the thickness of the surrounding epithelium.)

- ○ | **DISTAL CONVOLUTED TUBULE (DCT)** | : (confined to the renal CORTEX)

    - • *Simple cuboidal epithelium*

- Specialized for **selective secretion** and **reabsorption** of ions

- Function in conserving body fluids

    o   <u>CLASSES of NEPHRONS are DIVIDED ACCORDING to LOCATION:</u>

- **CORTICAL NEPHRONS** (85% of all nephrons) located almost entirely within the cortex

- **JUXTAMEDULLARY NEPHRONS** (15%): their renal corpuscles lie near the cortex-medulla junction

- **COLLECTING TUBULES** (or *collecting ducts*)

    o   Receive urine from several nephrons and run straight through the cortex into the deep medulla

    o   **PAPILLARY DUCTS:** adjacent collecting tubules join together to form these larger ducts, which empty into the minor calices through the renal papillae

    o   Walls consist of *simple cuboidal epithelium*, and thicken to become *simple columnar* in papillary ducts.

    o   Main function: **CONSERVE BODY FLUIDS**

    o   <u>During water conservation:</u> **ADH (*anti-diuretic hormone*)** is secreted by the posterior pituitary gland and **increases permeability of the collecting tubules and DCT to water.**

- **MICROSCOPIC BLOOD VESSELS** associated with the uriniferous tubules :

    o   **GLOMERULUS** (or *glomerular capillaries*): high-resistance vessels

    o   **PERITUBULAR CAPILLARIES:** arise from the efferent arterioles, draining the cortical glomerulus

- Surround the uriniferous tubules in the interstitial connective tissue of the renal cortex

- Low-pressure, porous capillaries adapted for absorption

    o   **VASA RECTA:** thin-walled looping vessels that surround the juxtamedullary nephron

- Run alongside the loops of Henle in the deepest part of the renal cortex

- Play a role in the kidney's **urine-concentrating mechanism**

- **JUXTAGLOMERULAR APPARATUS** :

  STRUCTURE THAT FUNCTIONS IN REGULATION OF BLOOD PRESSURE (BP)

  - Area of specialized contact between the 1st part of the DCT and the juxtaglomerular cells:

    - 1st part of the DCT contains **MACULA DENSA CELLS**, which act as CHEMORECEPTORS to **monitor solute concentration in the filtrate**

    - **JUXTAGLOMERULAR CELLS** – surround the afferent and efferent arterioles and act as MECHANORECEPTORS to **monitor BP**

      - **Secrete renin** (kidney hormone), which increases blood-solute concentration, blood volume and most importantly, increase BP

THE URETERS

- Slender tubes (~25 cm or 10 inches long) that carry urine from the kidneys to the bladder; continuation of the renal pelvis

- **HISTOLOGY:** 3 basic layers of the ureter walls
  (*DEEP/internal* to *SUPERFICIAL/external*)

  - **MUCOSA:** *transitional epithelium* (stretches when ureters fill with urine) + lamina propria (stretchy, fibroelastic connective tissue)

  - **MUSCULARIS:** inner *longitudinal* layer and outer *circular layer of smooth muscle*

    - A 3rd layer or external *longitudinal* layer appears in the inferior third of the ureter.

  - **ADVENTITIA:** typical connective tissue

- **Peristaltic** waves propel urine to the bladder.

URINARY BLADDER

- COLLAPSIBLE, MUSCULAR SAC that temporarily STORES and EXPELS URINE

- Lies anterior to the rectum in males; lies anterior to the vagina and inferior to the uterus in females

- Varying dimensions, depending on the state of distension

  - Maximum capacity is typically about 1 liter of urine.

- **FEATURES of the URINARY BLADDER:**

  - **POSTEROLATERAL ANGLES**, or ***ureteral openings*** (2): slit-like openings that receive the ureters

  - The superior surfaces of the urinary bladder are covered by a layer of peritoneum:

    o Several peritoneal folds assist in stabilizing the position of the urinary bladder

    o LATERAL UMBILICAL LIGAMENTS (2) pass along the sides of the bladder, and also assist in stabilizing bladder position

  - **URACHUS** (or ***median umbilical ligament***): fibrous band at the bladder's anterior angle

    o Extends from the anterior and superior border toward the umbilicus

    o Contains the closed remnant of an embryonic tube called *allantois*, which is a vestige of the umbilical arteries that supplied blood to the placenta during embryonic and fetal development

  - **INFERIOR ANGLE (NECK):** drains into the urethra

    o In males, the prostate gland lies directly inferior to bladder surrounding the urethra

  - **TRIGONE:** triangular region on the posterior wall of the bladder interior defined by openings for both ureters and the urethra

    o Its mucosa LACKS RUGAE (folds).

    o Smooth and very thick in appearance

    o Acts as a funnel that channels urine into the urethra when the bladder contracts

    o Of special clinical importance because infections tend to persist in this region

  - **HISTOLOGY** : 3 layers of the urinary bladder wall
    (*DEEP/internal* to *SUPERFICIAL/external*)

    o **MUCOSA:** transitional epithelium + lamina propria + submucosa

      • Lines the bladder interior

      • RUGAE in the mucosal lining disappear as the bladder stretches and fills with urine.

o **MUSCULARIS:** | **DETRUSOR MUSCLE** | consists of highly intermingled smooth muscle fibers arranged as follows:

- Inner *longitudinal* layer of smooth muscle

- Middle *circular* layer

- Outer *longitudinal* layer of smooth muscle

- **Contraction** of the DETRUSOR **compresses** the urinary bladder and **expels** its contents into the urethra.

o **FIBROUS ADVENTITIA:** outer connective tissue layer

- A layer of **serosa** (*visceral peritoneum*) covers the superior surface of the urinary bladder.

URETHRA

- Thin-walled tube that DRAINS URINE FROM THE BLADDER TO THE BODY EXTERIOR

- Comprised of smooth muscle and an inner mucosa; in males, the muscular layer becomes very thin toward the distal end of the **urethra** (pl. = ***urethrae***)

- The urethra extends from the NECK, or INFERIOR ANGLE of the URINARY BLADDER to the exterior.

- The male and female urethrae differ in length and in function.

  - **In the female urethra:**

    o Urethra is very short, about 3-5 cm (1-1.5 inches) from the bladder to the vestibule

    o The external urethral orifice (opening) is situated near the anterior wall of the vagina.

  - **In the male urethra:**

    o Urethra extends from the NECK of the bladder to the tip of the penis (18-20 cm, or 7-8 inches) o ⬚ Male urethra is subdivided into three portions ⬚ :

- **PROSTATIC URETHRA** – passes through the center of the prostate gland

- **MEMBRANOUS URETHRA** – includes a short segment that penetrates the urogenital diaphragm, which is the muscular floor of the pelvic cavity

- **SPONGY URETHRA** (or *penile urethra*) – extends from the distal border of the urogenital diaphragm to the external urethral orifice at the tip of the penis

- <u>**INTERNAL URETHRAL SPHINCTER:**</u> thickening of the detrusor muscle at the bladder-urethra junction

  - **INVOLUNTARY** sphincter of smooth muscle that keeps the urethra closed when urine is not being passed & prevents dribbling of urine between voidings/urination

- <u>**EXTERNAL URETHRAL SPHINCTER:**</u> surrounds urethra within the sheet of muscle called urogenital diaphragm; skeletal muscle

  - **VOLUNTARY** sphincter to inhibit urination until proper time

⬚ **MICTURITION** ⬚ (*voiding* or *urination*)

- The act of emptying the bladder is brought about by the **contraction of the detrusor muscle**, assisted by muscles of the abdominal wall, which contract to increase the intra-abdominal pressure.

<u>**URINE FLOWS through the RENAL TUBULE of the KIDNEY in the following order:**</u>

GLOMERULAR (or **BOWMAN'S**) **CAPSULE** → **PROXIMAL CONVOLUTED TUBULE** → **LOOP of HENLE** → **DISTAL CONVOLUTED TUBULE** → **COLLECTING DUCT or TUBULE** → **PAPILLARY DUCT** → **MINOR CALYX (CALICES)** → **MAJOR CALYX (CALICES)** → **RENAL PELVIS** → **URETER** → **URINARY BLADDER** → **URETHRA**

Chapter 25 – THE REPRODUCTIVE SYSTEM

OVERVIEW

- Both male and female reproductive systems' main function is to **produce offspring**.

- The primary sex organs are the **gonads** (*ovaries* or *testes*), which **produce gametes** (*ova* or *sperm*) **and sex hormones**.

- All other male and female reproductive organs are accessory organs.

THE MALE REPRODUCTIVE SYSTEM

- The PRIMARY SEX ORGANS, the **GONADS**, in the male reproductive system are the **TESTES** (singular = **testis**), which lie in the scrotum.

- Motile sperm travel to the body exterior from the testes through a system of **ACCESSORY reproductive** **DUCTS** in the following order:

  - The DUCTUS (or VAS) DEFERENS (*duct of the epididymis*)

  - The EJACULATORY DUCT

  - The URETHRA, which opens at the TIP of the PENIS

- Besides the accessory duct system, other male accessory reproductive organs, the **ACCESSORY GLANDS** (*empty their secretions into the sex ducts during ejaculation*), are:

  - The SEMINAL VESICLES

  - The PROSTATE GLAND

  - The BULBOURETHRAL GLANDS

- The **EXTERNAL GENITALIA** are the SCROTUM and the PENIS.

- **THE SCROTUM**

  - **STRUCTURE:** a SAC comprised of SKIN and SUBCUTANEOUS TISSUE (SUPERFICIAL FASCIA), which contains the testes or testicles

  - **LOCATION of the SCROTUM:** SUPERFICIAL and inferiorly EXTERNAL to the abdominopelvic cavity at the root of the penis

  - **FUNCTION of the SCROTUM:** responds to changes in external temperature

    o Under cold conditions, the testes are pulled up toward the warm body wall and the scrotal skin wrinkles to increase its thickness and reduce heat loss.

    o These effects are accomplished through the actions of the **DARTOS MUSCLE** (*for skin wrinkling*) and the **CREMASTER MUSCLES** (*for elevating testes*)

- **THE TESTES** (sing. = *testis*)

  - **GROSS ANATOMY of the TESTES**

    o Each testis is partially surrounded by:

      - a <u>superficial</u> 3-layered serous sac called the **TUNICA VAGINALIS** (*derived from peritoneum*) and;

      - a <u>deeper</u> layer called the **TUNICA ALBUGINEA** (*fibrous capsule* of the testis).

    o Each testis is divided into many **LOBULES** that contain sperm-producing **SEMINIFEROUS TUBULES**, which are separated by the *tunica albuginea*.

    o The thick epithelium of the SEMINIFEROUS TUBULE consists of the following

      - **SPERMATOGENIC CELLS** (spermatogonia, primary and secondary spermatocytes, and spermatids), which are embedded in;

      - **SUSTENTACULAR** (or **SERTOLI**) **CELLS** (or *nurse cells*) – cells that are attached to the basal lamina of the seminiferous tubule capsule

FUNCTIONS of SERTOLI CELLS (or *nurse cells*):

- Form the **BLOOD-TESTIS BARRIER**

- **Nourish** the **spermatogenic cells**, and **move them toward the lumen**

- They also <u>secrete testicular fluid</u> (for sperm transport), <u>androgen-binding protein</u> (ABP), and the hormone <u>inhibin</u>.

○ Posteriorly, the seminiferous tubules of each lobule converge to form a **TUBULUS RECTUS** (plural = *tubuli recti*), **or STRAIGHT TUBULE**, which conveys sperm into the RETE TESTES (*plural form*).

○ **RETE TESTIS** (*singular form*): complex network of tiny branching tubes, which empty into **EFFERENT DUCTULES** that enter the **EPIDIDYMIS**.

○ **In the seminiferous tubules, spermatogenic cells** move toward the tubule lumen as they <u>differentiate into sperm</u> by a process called **SPERMATOGENESIS**.

- **SPERMIOGENESIS:** the differentiation or maturation process in which a spermatid becomes a mature spermatozoa

  - **SPERMIATION:** the process in which a spermatozoon becomes detached from the sertoli cell and enters the lumen of the seminiferous tubule; marks the end of spermiogenesis

○ **MYOID CELLS** are smooth-muscle-like cells that <u>surround the seminiferous tubules</u> and HELP TO SQUEEZE SPERM through the TUBULES and out of the testes when they contract rhythmically.

○ **INTERSTITIAL (or LEYDIG) CELLS** (*oval cells in the connective tissue between seminiferous tubules*) secrete the following male sex hormones:

- **ANDROGENS (*mostly testosterone*)**, under the influence of **luteinizing hormone (LH)**, which is secreted from the pituitary gland

- Testosterone also maintains all male sex characteristics and sex organs.

- ## THE ACCESSORY REPRODUCTIVE DUCT SYSTEM in MALES

  - From the seminiferous tubules, motile sperm travel through the following structures in order:

    o TUBULI RECTI and RETE TESTES, then out of the testes through;

    o the **EFFERENT DUCTULES** into;

    o the **DUCT of the EPIDIDYMIS**

  - The comma-shaped **EPIDIDYMIS** hugs the posterolateral surface of the testis.

    o The **DUCT of the EPIDIDYMIS or DUCTUS (VAS) DEFERENS** (*lined with CILIATED pseudostratified columnar epithelium*) is where SPERM GAIN THE ABILITY TO SWIM and FERTILIZE.

    o EJACULATION begins with the CONTRACTION of SMOOTH MUSCLE in the DUCT of the EPIDIDYMIS.

  - The **DUCTUS (or VAS) DEFERENS** extends from the epididymis to the ejaculatory duct in the pelvic cavity.

    o During ejaculation, the thick layers of smooth muscle in its wall PROPEL SPERM into the URETHRA by PERISTALSIS.

  - The fascia-covered **SPERMATIC CORD** contains the *ductus deferens* and the **testicular vessels and nerves**.

  - The **URETHRA** (runs from the bladder to the tip of the penis) conducts semen and urine to the body exterior.

    Consists of three parts:
    o **PROSTATIC** URETHRA in the prostate gland

    o **MEMBRANOUS** URETHRA in the urogenital diaphragm

    o **SPONGY (or PENILE)** URETHRA in the penis

- ## ACCESSORY GLANDS

  - **Produce the bulk of the semen**, which is comprised of motile sperm and the secretions of the accessory glands and accessory ducts

- **SEMINAL VESICLES** : long, pouched glands posterior to the bladder

  o They **secrete a sugar-rich fluid** that constitutes 60% of the ejaculate.

- **PROSTATE GLAND** (surround the prostatic urethra): a group of **compound glands embedded in a fibromuscular stroma** (*smooth muscle*), which contracts during ejaculation to squeeze the prostatic secretion into the urethra

  o **Its secretion constitutes ⅓ volume of semen and is a milky fluid** that contains various substances to enhance sperm motility, and enzymes that help to clot, then to liquefy, the semen.

- **BULBOURETHRAL GLANDS** (located in the urogenital diaphragm): pea-sized glands, which **secrete mucus into the urethra before ejaculation** to neutralize traces of acidic urine and to lubricate the urethra for the passage of semen.

- **THE PENIS**

  - Male organ of sexual intercourse; considered as external genitalia, along with the scrotum

  - **FUNCTION:** delivers sperm into the female reproductive tract

  - **STRUCTURE:**

    o The penis is divided into three regions:

    - the **ROOT** of the penis – the fixed portion that attaches the penis to the rami of the ischia, within the urogenital triangle immediately inferior to the pubic symphysis

    - the **SHAFT (BODY)** of the penis – the tubular, movable portion that is comprised of masses of *erectile tissue* (*corpora*)

    - the **GLANS** of the penis – the expanded distal portion that surrounds the external urethral orifice

      - Covered by a fold of skin, called the **PREPUCE**, or **FORESKIN**

○ Its main nerves and vessels lie <u>dorsally</u> in the midline.

○ Internally, the penis contains the SPONGY URETHRA and 3 long, cylindrical bodies (*corpora*; singular = *corpus*) of erectile tissue.

- **CORPUS SPONGIOSUM** surrounding the spongy urethra (forms the GLANS PENIS DISTALLY and a part of the root called the BULB OF THE PENIS PROXIMALLY) and;

- **TWO (paired) dorsal** **CORPORA CAVERNOSA** located lateral to the corpus spongiosum

  - They diverge at their bases, forming the **CRURA** (singular = *crus*) **of the penis**, each of which is bound to the ramus of the ischium via tough connective tissue ligaments.

  - Each *corpus cavernosum* contains a **DEEP ARTERY of the PENIS**.

  - Engorgement of these bodies with blood causes ERECTION.

  - SEMEN RELEASE involves a two-step process :

    ○ **EMISSION:** process in which the sympathetic nervous system coordinates peristaltic contractions that mix the fluid components of the semen within the male reproductive tract

    ○ **EJACULATION:** process in which powerful rhythmic contractions begin in the ***ischiocavernosus*** and ***bulbospongiosus*** *muscles* of the pelvic floor to <u>stiffen the penis</u> and <u>push semen toward the external urethral orifice</u>, respectively

- **THE MALE PERINEUM**

  - The diamond-shaped perineum contains the scrotum, the root of the penis and the anus.

THE FEMALE REPRODUCTIVE SYSTEM

- The PRIMARY SEX ORGANS, the **GONADS**, are the **OVARIES** , which <u>produce the gametes</u> (***ova***; singular = ***ovum***) and sex hormones.

- The ACCESSORY DUCTS include:

  - The UTERINE TUBES (where fertilization typically occurs)

  - The UTERUS (where the embryo develops)

  - The VAGINA (acts as a birth canal and receives the penis during sexual intercourse).

- The EXTERNAL GENITALIA are referred to as the VULVA or PUDENDUM, and include the following structures:

  - The MONS PUBIS
  - The LABIA
  - The CLITORIS
  - STRUCTURES ASSOCIATED with the VESTIBULE (the space bounded by the external genitalia)

- The MAMMARY GLANDS are actually part of the integumentary system, but are considered in this chapter because of their REPRODUCTIVE FUNCTION of NOURISHING the INFANT.

- **ADDITIONAL FUNCTIONS:**

  - The female reproductive system also **houses, nourishes, protects, and delivers a developing infant.**

  - It also **undergoes a menstrual cycle** of about 28 days.

- THE OVARIES

  - **GROSS ANATOMY:**

    - The almond-shaped ovaries lie on the lateral walls of the pelvic cavity, flank the uterus on each side and are **suspended by various mesenteries and ligaments.**

      - The **MESOVARIUM** (*mesentery of the ovary*) is part of a fold of peritoneum called the **BROAD LIGAMENT**, which hangs like a tent from the uterus and uterine tubes and also continues as the **SUSPENSORY LIGAMENT of the OVARY** and its continuation, the **ROUND LIGAMENT of the UTERUS.**

323

- The OVARIAN LIGAMENT, a distinct fibrous band enclosed within the broad ligament, anchors the ovary to the uterus medially.

○ | Each ovary is divided into two regions | :

- the **OUTER CORTEX** – the region that houses the developing gametes, called OOCYTES, which occur within saclike multicellular structures called FOLLICLES
  - These follicles enlarge substantially as they mature.

- the **INNER MEDULLA** – the region comprised of loose connective tissue containing the MAIN OVARIAN BLOOD VESSEL, LYMPH VESSELS and NERVES

○ The ovaries of a newborn female contain many thousands of ***primordial follicles***, each of which consists of an *oocyte* surrounded by a layer of flat *follicular cells*.

■ | **The OVARIAN CYCLE** (the MENSTRUAL CYCLE as it relates to the OVARY). |

The ovarian cycle has three successive phases:

○ The **FOLLICULAR PHASE** (*Days 1-14*):

- 6-12 follicles start maturing

- Generally, only one follicle per month completes the maturation process.

○ **OVULATION** (*Day 14*):

- Upon stimulation by ***luteinizing hormone (LH)***, the oocyte is released from the ovary into the peritoneal cavity

- Involves a weakening and rupture of the follicle wall followed by violent muscular contraction of the *external theca cells*

○ The **LUTEAL PHASE** (*Days 15-28*):

- The ruptured follicle remaining in the ovary becomes a wavy **CORPUS LUTEUM that secretes *progesterone*** (acts on mucosa of uterus, signaling it to prepare for implantation of an embryo) **and *estrogens*.**

- **If fertilization does not occur,** the corpus luteum degenerates in about 2 weeks into a CORPUS ALBICANS (scar), which remains in the ovary for several months and shrinks until it is finally phagocytosed by macrophages.

- **OOGENESIS** - the production of the female gametes (*ova*)

    o   Starts before birth and takes decades to complete

    o   The stem cells (**OOGONIA**) appear in the ovarian follicles of the fetus.

    o   *Primary oocytes* stay in **meiosis I** until <u>ovulation</u> occurs years later.

    o   Each *secondary oocyte* then stays in **meiosis II** <u>until a sperm penetrates it</u>.

- **THE UTERINE TUBES** (also called FALLOPIAN TUBES or OVIDUCTS)

    - <u>FUNCTION</u>: receive the ovulated oocyte and provide a site for fertilization

    - <u>STRUCTURE</u>:

        o   Each uterine tube extends from an ovary laterally to the uterus medially.

        o   Its **regions**, from **lateral to medial**, are:

            • the **INFUNDIBULUM**

            • the **AMPULLA**

            • the **ISTHMUS**

        o   The ciliated, finger-like projections called FIMBRIAE (drape over the ovary) extending from the infundibulum create currents that help draw an ovulated oocyte into the uterine tube.

    - <u>HISTOLOGY</u> of the wall of the uterine tube:

        o   a *muscularis layer* comprised of *smooth muscle*

        o   a folded inner *mucosa* with a *simple columnar epithelium*

Both the smooth muscle and ciliated epithelial cells help propel the oocyte toward the uterus.

- **THE UTERUS** ("womb")

  - **LOCATION and STRUCTURE**

    - Located in the pelvic cavity, <u>anterior to the rectum</u> and <u>posterosuperior to the bladder</u>

    - The hollow uterus, shaped like an inverted pear, has four regions:

      - a **FUNDUS** - rounded region superior to the entrance of the uterine tubes

      - a **BODY**

      - an **ISTHMUS** - slightly narrowed region inferior to the body

      - a **CERVIX** - narrow neck inferior to isthmus, projects into the vagina

        - *Cervical glands* fill the **CERVICAL CANAL** with a bacteria-blocking mucus.

        - The cervical canal communicates with the vagina inferiorly via the **EXTERNAL OS** and with the cavity of the uterine body superiorly via the **INTERNAL OS**.

  - **LIGAMENTS and MESENTERIES of the uterus**

    - The uterus is anchored to the lateral pelvic walls by the **MESOMETRIUM**, which is the largest division of the broad ligament.

    - It is also supported by the **BROAD, LATERAL CERVICAL, and ROUND LIGAMENTS**.

    - **Most support**, however, comes from the **muscles of the pelvic floor**.

  - **UTERINE WALL**

    - <u>The uterine wall consists of three basic layers (*outer to external*):</u>

      - the **PERIMETRIUM** - the <u>outer</u> *serous membrane*, which is the peritoneum

      - the **MYOMETRIUM** – a <u>middle</u> bulky layer consisting of interlacing bundles of *smooth muscle*

        - Functions to squeeze the baby out during childbirth

- the **ENDOMETRIUM** – an <u>inner</u> thick *mucosa* that consists of *simple columnar epithelium containing secretory and ciliated cells*

> **The endometrium consists of two layers, or *strata* (sing. = *stratum*)** :

- **STRATUM FUNCTIONALIS (or FUNCTIONAL LAYER)** – inner layer closest to the uterine cavity

  - Contains most of the uterine glands and contributes most of the endometrial thickness

  - Undergoes cyclic changes in response to varying levels of ovarian hormones and sloughs off each month (except in pregnancy)

- **STRATUM BASALIS (or BASILAR LAYER)** – outer layer that is adjacent to the myometrium

  - It attaches the endometrium to the myometrium.

  Contains the terminal branches of the tubular glands

  - It is not shed, but it replenishes the *stratum functionalis*

- ARTERIAL SUPPLY of the UTERINE WALL :

  - Branches of the uterine arteries form **ARCUATE ARTERIES** that encircle the endometrium.

  - **RADIAL ARTERIES** branch from the arcuate arteries and <u>supply the straight arteries and the spiral arteries.</u>

  - **STRAIGHT ARTERIES** supply the *stratum basalis* or basilar layer of the endometrium.

  - **SPIRAL ARTERIES** supply the *stratum functionalis* or functional layer.

- **The UTERINE CYCLE is the MENSTRUAL CYCLE** (as it relates to the ENDOMETRIUM).

  - The uterine cycle averages 28 days in length, but it can range from 21 – 35 days in normal individuals.

  - These endometrial phases are closely coordinated with the phases of the ovarian cycle.

- The phases occur in response to hormones associated with the regulation of the ovarian cycle.

| The uterine cycle has three successive phases. |

---

o <u>The **MENSTRUAL** PHASE, or **MENSES**</u>:

- It is the onset of the uterine cycle.

- Marked by the <u>destruction of the functional layer</u> of the endometrium

o <u>The **PROLIFERATIVE** PHASE</u>:

- The <u>functional layer undergoes repair and thickens</u>.

  - The epithelial cells of the uterine glands multiply and spread across the endometrial surface, restoring the integrity of the uterine epithelium.

- The end result is complete restoration of the functional layer.

o <u>The **SECRETORY** PHASE</u>:

- The endometrial glands enlarge and increase their rates of secretion.

- The <u>arteries elongate and spiral through the tissues</u> of the functional layer.

o The menstrual and proliferative phases are a shedding and then a rebuilding of the endometrium in the 2 weeks before ovulation.

o The third phase prepares the endometrium to receive an embryo in the 2 weeks after ovulation.

o **MENARCHE:** the first uterine cycle at puberty, which typically occurs at age 11-12

o **MENOPAUSE:** the last uterine cycle, which typically occurs at age 45-55

- **THE VAGINA** ("birth canal")

  - A highly distensible muscular tube that runs from the cervix of the uterus to the body exterior at the vestibule

  - Average length is 7.5 – 9 cm (3 – 3.5 inches)

  - **PRIMARY BLOOD SUPPLY**

    - VAGINAL BRANCHES of the internal iliac (or UTERINE) ARTERIES and VEINS

  - **LOCATION of the VAGINA**

    - Lies inferior to the uterus

    - Anterior and parallel to the rectum

    - Posterior to the urethra and bladder

  - **FUNCTIONS of the VAGINA**

    - Receives the penis and semen during intercourse

    - Acts as the lower portion of the birth canal

    - Serves as a passageway for the elimination of menstrual fluids

  - HISTOLOGY of the VAGINAL WALL

    - OUTER **ADVENTITIA** – fibrous connective tissue layer covering the vagina

      - <u>Exception</u>: SEROSA that is continuous with the pelvic peritoneum covers the portion of the vagina adjacent to the uterus.

    - MIDDLE **MUSCULARIS** layer – layers of smooth muscle fibers arranged in circular and longitudinal bundles that are continuous with the uterine myometrium

    - INNER **MUCOSA** - consists of an elastic lamina propria (contains blood vessels, nerves, and lymph nodes) and a stratified squamous epithelium

- In the relaxed state, this layer is thrown into folds, called **RUGAE**.

- The VAGINAL LUMEN is **ACIDIC.**

  o The vaginal lining contains a **normal population of resident bacteria**, which are supported by the nutrients found in the cervical mucus.

  o The acidic environment is **due to the metabolic activity of the bacteria.**

  o The acidity **restricts the growth of many pathogenic organisms and inhibits sperm motility**.

    - Consequently, buffers in seminal fluid are important for successful fertilization.

- Near the vaginal orifice, the inner layer of mucosa elaborates to form an incomplete diaphragm called the **HYMEN**, which separates the vagina from the vestibule.

  o The hymen is vascular and tends to bleed when ruptured during the first sexual intercourse.

  o However, this is not always the case because its durability varies and may be ruptured by other means such as sports activity, insertion of a tampon, or a pelvic examination.

- **VAGINAL FORNIX** : a shallow ring-like recess around the tip of the cervix in the superior vagina.

  o **CERVICAL PROTRUSION** – area where the cervix projects into the vaginal canal, at the proximal end of the vagina

- **THE EXTERNAL GENITALIA**

The female external genitalia (also called **VULVA** or **PUDENDUM**) include the following:

- **MONS PUBIS** : fatty, rounded pad overlying the pubic symphysis

- The **LABIA MAJORA** :

330

o   HOMOLOGUES of the SCROTUM, derived from same embryonic structure

o   Two long, hair-covered, fatty skin folds extending posteriorly from the *mons pubis*

- The   **LABIA MINORA** :

  o   Two thin, hairless folds of skin enclosed by the *labia majora*

- The   **CLITORIS** :

  o   Located anterior to the urethral opening and projects into the vestibule

  o   A protruding structure composed largely of erectile tissue that is HOMOLOGOUS with the CORPORA CAVERNOSA in males

    •   Hooded by a fold of skin, the **PREPUCE**, or **HOOD**, of the CLITORIS, which arises from extensions of the *labia minora* that encircle the body of the clitoris

    •   **GLANS** – small erectile tissue area that sits atop the clitoris

    •   **BODY** of the CLITORIS contains **paired CORPORA CAVERNOSA**

    •   <u>During sexual stimulation</u>, the homologous **BULBS of the VESTIBULE**, or **VESTIBULAR BULBS** (*erectile bodies*) <u>ENGORGE with BLOOD and may help grip the penis within the vagina</u>.

      ▪   The vestibular bulbs are **homologous to the corpus spongiosum** in males.

      ▪   They lie along each side of the vaginal orifice and directly deep to the *bulbospongiosus muscle*.

- The **VESTIBULE**

  o   Enclosed by the labia minora

  o   Houses the VAGINAL (posterior) ORIFICE and the URETHRAL (anterior) ORIFICE

  o   The mucus-secreting *greater vestibular glands* (or *Bartholin's glands*) and the bulbs of the vestibule (erectile bodies) lie just deep to the labia.

- **GLANDS** associated with the **VESTIBULE**

  - **PARAURETHRAL GLANDS (or SKENE'S GLANDS)** discharge <u>lubricating fluid</u> into the urethra near the external urethral orifice.

  - **LESSER VESTIBULAR GLANDS**

    - A variable number of these glands discharge their secretions onto the <u>exposed surface of the vestibule</u>, to keep it <u>moistened</u>.

  - **GREATER VESTIBULAR GLANDS**

    - During arousal, these glands discharge their secretions (via a pair of ducts) into the vestibule near the posterolateral margins of the vaginal entrance.

    - These <u>mucous glands</u> resemble the bulbourethral glands in the male reproductive system.

- **THE FEMALE PERINEUM**

  - Diamond-shaped region between the pubic arch anteriorly, the coccyx posteriorly and the ischial tuberosities laterally

  - The central tendon of the perineum lies just posterior to the vaginal orifice and the ***fourchette*** (or ***frenulum of the labia***) where the right and left *labia minora* come together to form a ridge.

- **THE MAMMARY GLANDS**

  - The mammary glands develop from the skin of the embryonic milk lines.

  - Technically, these glands are a part of the integumentary system, but functionally belong to the reproductive system.

  - Lie in the *subcutaneous layer* beneath the skin of the chest

  - Function as the site of **LACTATION** (*milk production*)

- ○ Internally, each breast consists of 15 – 25 lobes (*compound alveolar glands*) that secrete milk.

- The **LOBES** (subdivided into **LOBULES** and **ALVEOLI**) are separated by adipose tissue and by supportive suspensory ligaments.

- The <u>SECRETORY LOBULES</u> consists of:

  - ○ **LACTIFEROUS DUCT** – convergence (union) of ducts leaving the lobules

    - • **Breast cancers** usually arise from the lactiferous duct system.

  - ○ **LACTIFEROUS SINUS** – expansion of ducts leaving the lobules, near the nipple

    - • Ducts of underlying mammary glands open onto the body surface at the **NIPPLE**

- Branches of the **INTERNAL THORACIC ARTERY** supply each breast.

- The full glandular structure of the breast does not <u>develop</u> until the second half of pregnancy, <u>under the influence of hormones</u>.

  - ○ **PROLACTIN (PRL)** – secreted from the anterior pituitary

  - ○ **GROWTH HORMONE (GH)** – from the anterior pituitary

  - ○ **HUMAN PLACENTAL LACTOGEN (HPL)** – from the placenta

PRACTICE QUESTIONS 6:
The Respiratory, Digestive, Urinary, and Reproductive Systems

1. The basic functional unit of the liver is the:
   a. hepatocyte.
   b. Kuppfer cell.
   c. lobule.
   d. falciform ligament.
   e. glomerulus.

2. Which of the following is *not* a function of the respiratory system?
   a. protection from dehydration by filtering water out of the air
   b. movement of air to and from the exchange surfaces
   c. production of sound
   d. providing an extensive surface area for gas exchange
   e. All of the above are functions of the respiratory system.

3. Which of the following is a small intestine feature that increases the surface area for digestion and absorption?
   a. cilia
   b. microvilli
   c. haustra
   d. teniae coli
   e. rugae

4. Which of the following absorbs digested fats?
   a. absorptive cells
   b. lacteals
   c. enteroendocrine cells
   d. rugae
   e. intestinal crypts

5. Contraction of the cremaster muscle:
   a. propels sperm through the urethra.
   b. moves sperm through the ductus deferens.
   c. produces an erection.
   d. moves the testis closer to the body cavity.
   e. B and C

6. An hepatic (portal) triad consists of branches of the:
   a. central vein, hepatic vein, and bile duct.

b.   hepatic portal vein, central vein, and hepatic artery.

c.   hepatic portal vein, hepatic artery, and hepatic vein.

d.   hepatic portal vein, hepatic artery, and bile duct.

e.   hepatic artery, central vein, and bile duct.

7.   The ureters:

a.   are not retroperitoneal.

b.   take exactly the same path to the bladder in men and women.

c.   have a layer of transitional epithelium.

d.   have specialized subdivisions called the urethrae.

e.   A and B

8.   The process whereby ovum production occurs is:

a.   atresia.

b.   oogenesis.

c.   triggered by completely different hormones than those of the male that initiate spermatogenesis.

d.   continuous throughout the life of the individual.

e.   None of the above.

9.   Compared with the right primary bronchus, the left primary bronchus is:

a.   more prone to blockage.

b.   steeper.

c.   less resistant to airflow.

d.   All of the above.

e.   None of the above.

10. Which of the following is a function of Sertoli cells (*nurse cells*)?

a.   nourish the spermatogenic cells

b.   secrete androgen-binding protein (ABP)

c.   form the blood-testis barrier

d.   A and C

e.   All of the above.

11. Which cells produce surfactant?

a.   alveolar macrophages

b.   Type I cells

c.   endothelial cells

d.   Type II cells

e.   B and D

12. The left lung:

a.   has three lobes.

b.   has a cardiac notch.
c.   is supplied entirely by a secondary bronchus.
d.   receives oxygenated blood from the heart.
e.   All of the above.

13. Which arteries directly supply the basilar layer of the endometrium?
    a.   radial arteries
    b.   straight arteries
    c.   testicular arteries
    d.   spiral arteries
    e.   genitofemoral arteries

14. The mechanism of urine production in the uriniferous tubule, a process in which most of the *nutrients, water and essential ions are reclaimed from the filtrate* is called:
    a.   micturition.
    b.   secretion.
    c.   reabsorption.
    d.   filtration.
    e.   None of the above.

15. Which of the following is located on the posterior wall of the bladder interior, defined by the openings for both ureters and for the urethra?
    a.   trigone
    b.   inferior angle (neck)
    c.   internal urethral sphincter
    d.   posterolateral angle
    e.   urachus

16. The layer, which adheres directly to the kidney's surface is the:
    a.   pararenal fat.
    b.   adipose capsule.
    c.   renal fascia.
    d.   fibrous capsule.
    e.   perirenal fat.

17. Which female reproductive organ is homologous to the *corpus spongiosum*?
    a.   vaginal fornix
    b.   vestibular bulbs
    c.   clitoris
    d.   labia minora
    e.   labia majora

18. Blood exits the glomerulus via the:
    a. renal artery.
    b. efferent arteriole.
    c. afferent arteriole.
    d. lobar artery.
    e. segmental artery.

19. Which structure(s) marks the boundary between the superior nasopharynx and the rest of the pharynx?
    a. paranasal sinuses
    b. hard palate
    c. nasal conchae
    d. inferior meatus
    e. soft palate

20. Which layer of the tracheal wall contains seromucous glands?
    a. adventitia
    b. submucosa
    c. muscularis externa
    d. serosa
    e. mucosa

21. Which structure(s) is/are found in the large intestine?
    a. teniae coli
    b. plicae circulares
    c. rugae
    d. villi
    e. hepatopancreatic ampulla

22. Which structure leads directly from the gallbladder toward the porta hepatis?
    a. common hepatic duct
    b. right hepatic duct
    c. cystic duct
    d. left hepatic duct
    e. common bile duct

23. The pancreatic and pancreato-duodenal arteries are major branches from the:
    a. splenic, gastric, and superior mesenteric arteries.
    b. inferior mesenteric, common hepatic, and splenic arteries.
    c. pancreatic, gastric, and splenic arteries.
    d. common hepatic, superior mesenteric, and splenic arteries.
    e. C and D

24. Which layer appears in the inferior third of the ureters?
    a. external longitudinal smooth muscle layer
    b. mucosa
    c. external circular smooth muscle layer
    d. adventitia
    e. fibroelastic lamina propria

25. Compression of which structure(s) compresses the urinary bladder, expelling its contents into the urethra?
    a. detrusor muscle
    b. median umbilical ligament
    c. cremaster muscle
    d. bulbospongiosus muscle
    e. lateral umbilical ligament

26. Which phase of the ovarian cycle involves a weakening and rupture of the follicle wall?
    a. ovulation
    b. proliferative phase
    c. luteal phase
    d. secretory phase
    e. follicular phase

APPENDIX A
RECOMMENDED REFERENCE TEXTS

The following is a list of recommended textbooks, which were used for reference in compiling and verifying facts in this compendium.

Human Anatomy, *Fifth Edition*, by **Elaine N. Marieb**, Pearson Benjamin Cummings Publishing, 2007

Human Anatomy, *Sixth Edition*, by **Frederic H. Martini**, Pearson Benjamin Cummings Publishing, 2008

Essential Clinical Anatomy, *Second Edition* (or *Third Edition*), **Keith L. Moore**, Lipppincott Williams & Wilkins, 2002 (2006)

Human Anatomy, *Second Edition*, **Kenneth S. Saladin**, McGraw-Hill Science Engineering, 2007

Principles of Human Anatomy, *Eleventh Edition*, **Gerard J. Tortora**, John Wiley & Sons, Inc., 2008

Human Anatomy, *Second Edition*, **Michael P. McKinley** and **Valerie D. O'Loughlin**, McGraw-Hill Science Engineering, 2007

Cranial Nerves Anatomy and Clinical Comments, **Linda Wilson-Pauwels**, B.C. Decker, Inc., 1988

Cranial Nerves in Health and Disease, **Linda Wilson-Pauwels**, B.C. Decker, Inc., 2001

Clinically Oriented Anatomy, *Fourth Edition*, **Keith L. Moore**, Lippincott Williams & Wilkins, 1999 (*Sixth Edition* - published February 2009)

APPENDIX B
ANSWERS to PRACTICE QUESTIONS

PRACTICE QUESTIONS 1:

1. A	9. A	17. A	25. C
2. C	10. E	18. D	26. A
3. C	11. E	19. E	27. D
4. B	12. C	20. D	28. A
5. A	13. D	21. E	29. C
6. E	14. B	22. E	30. D
7. A	15. E	23. B	
8. B	16. E	24. E	

PRACTICE QUESTIONS 2:

1. B	8. A	15. A	22. B	29. D	36. D
2. D	9. C	16. C	23. E	30. B	
3. A	10. D	17. D	24. D	31. C	
4. E	11. A	18. D	25. A	32. B	
5. B	12. D	19. D	26. A	33. E	
6. C	13. E	20. C	27. A	34. A	
7. E	14. D	21. D	28. E	35. D	

PRACTICE QUESTIONS 3:

1. D	6. B	11. B	16. A	21. E	26. E
2. A	7. E	12. C	17. E	22. E	27. B
3. E	8. C	13. B	18. B	23. B	28. E
4. B	9. E	14. B	19. C	24. A	
5. D	10. D	15. C	20. C	25. C	

PRACTICE QUESTIONS 4:

1. E	8. E	15. B	22. A	29. E
2. E	9. B	16. D	23. B	
3. B	10. E	17. C	24. E	
4. C	11. A	18. D	25. A	
5. A	12. C	19. B	26. C	
6. B	13. D	20. D	27. B	
7. C	14. C	21. C	28. D	

ANSWERS to PRACTICE QUESTIONS

PRACTICE QUESTIONS 5:

1. C	7. D	13. B	19. A
2. D	8. D	14. C	20. C
3. E	9. C	15. E	21. D
4. B	10. C	16. E	22. A
5. A	11. A	17. D	23. B
6. D	12. A	18. B	24. A

PRACTICE QUESTIONS 6:

1. C	7. C	13. B	19. E	25. A
2. A	8. B	14. C	20. B	26. C
3. B	9. E	15. A	21. A	
4. B	10. E	16. D	22. C	
5. D	11. D	17. B	23. D	
6. D	12. B	18. B	24. A	

CPSIA information can be obtained at www.ICGtesting.com
Printed in the USA
238122LV00001B/5/P

9 781438 986487